D1272220

ROOT CAUSE ANALYSIS

Improving Performance for Bottom-Line Results

Second Edition

PLANT ENGINEERING SERIES

Series Editor
Robert J. Latino
Vice President of Strategic Development
Reliability Center, Inc.
Hopewell, Virginia

FORTHCOMING TITLES

Results-Oriented Maintenance Management
Christer S. Idhammar

Total Equipment Asset Management
S. Bradley Peterson

CMMS: A Time Saving Implementation Process
Daryl Brett Mather

PUBLISHED TITLES

**Root Cause Analysis: Improving Performance for
Bottom Line Results**
Robert J. Latino and Kenneth C. Latino

ROOT CAUSE ANALYSIS

Improving Performance for Bottom-Line Results

Second Edition

Robert J. Latino
Reliability Center, Inc.
Hopewell, Virginia

Kenneth C. Latino
Meridium, Inc.
Roanoke, Virginia

CRC PRESS

Boca Raton   London   New York   Washington, D.C.

Library of Congress Cataloging-in-Publication Data

Latino, Robert J.
   Root cause analysis : improving performance for bottom line results / Robert J. Latino.
Kenneth C. Latino.-- 2nd ed.
      p. cm. -- (Plant engineering series)
   Includes bibliographical references and index.
   ISBN 0-8493-1318-X (alk. paper)
      1. Quality control--Data processing. 2. Industrial accidents--Investigation. 3. Critical
incident technique. I. Latino, Kenneth C. II. Title. III. Plant engineering series (Boca
Raton, Fla.) ·

TS156 .L368 2002
658.5'62--dc21

                                                                              2002017538

This book contains information obtained from authentic and highly regarded sources. Reprinted material is quoted with permission, and sources are indicated. A wide variety of references are listed. Reasonable efforts have been made to publish reliable data and information, but the author and the publisher cannot assume responsibility for the validity of all materials or for the consequences of their use.

Visit the CRC Press Web site at www.crcpress.com

© 2002 by CRC Press LLC

No claim to original U.S. Government works
International Standard Book Number 0-8493-1318-X
Library of Congress Card Number 2002017538
Printed in the United States of America 1 2 3 4 5 6 7 8 9 0
Printed on acid-free paper

Preface

Because technology is outpacing humanity's ability to keep up with it, we are faced with a growing amount of losses. Whether losses are measured in terms of units produced, quality of healthcare, customer satisfaction, etc., the common denominator of why our numbers do not measure up to expectations is the human being.

As corporations continue to reengineer or downsize to "trim the fat," they raise the bar on their production expectations. Now we are faced with a paradox of having fewer people with more tasks to accomplish. How can this be done? The answer, and what we believe corporate America is trying to do, is reinvent the way we do work. Doing things the way we have always done them will no longer obtain the goal. Therefore, we must think outside the proverbial box and question, why we do the things we do and the way that we do them.

As complex as it seems in our respective environments, the answer is quite simple when we look at the big picture. Corporations set earnings expectations, plants set production goals, or hospitals set expected profit margins; whatever the case, they all set the bar at a certain level. Once that bar is set, all plans revolve around it. The dilemma that we all face is where are we now compared to the bar? The distance between our actual situation and where the corporation would like us to be is the "Gap." The gap is composed of various undesirable outcomes, failures, incidents, events, etc. What they all have in common is that they are siphoning money out of the corporate engine. These events are costing organizations a fortune, and most organizations do not even know what or where these events are located. Consider this thought, "Why does a maintenance budget exist?" It exists to repair both expected and unexpected events. For example, in a manufacturing plant when chronic type failures occur on a daily or weekly basis, we tend to become very good at repairing them. They happen so often that we do not see them as a failure any more but as part of the job. When this occurs, it is retired to the pasture of the maintenance budget and accepted as "a cost of doing business." In essence, we will attain our goals in spite of these events. This is a costly paradigm that is generally worth tens and hundreds of millions of dollars in accepted losses.

What if we decided we were not going to accept these small, chronic issues any more? What if we set out to identify all the events in the gap? What if we dedicated ourselves to understanding why error occurs and how to prevent it? What if we were to eliminate chronic failure from occurring? It is difficult for most to envision because it seems like it is such a distant goal. Our people have the knowledge and the power to move our organizations to heights never imaginable, if we would just believe in them and let them soar. This text is an effort for both management and analysts to help them fully understand why things do not always work out as planned and how to use root cause analysis to make sure they do not happen again.

Robert J. Latino
Kenneth C. Latino

About the Authors

Robert J. Latino is Vice President of Strategic Development for Reliability Center, Inc. (RCI). RCI is a Reliability Engineering firm specializing in improving equipment, process, and human reliability since 1972. Mr. Latino received his bachelor's degree in business administration and management from Virginia Commonwealth University. He has a special interest in the theory of human error as applied to root cause analysis (RCA). Mr. Latino has been a practitioner of RCA in the field with his clientele as well as an educator for 17 years.

Mr. Latino is an author of RCI's Root Cause Analysis Methods© course and coauthor of Problem Solving Methods (PSM)© for Field Personnel. Mr. Latino has been published in numerous trade magazines on the topic of RCA and is also a frequent speaker on the topic at trade conferences.

Kenneth C. Latino has a bachelor's of science degree in computerized information systems. He began his career developing and maintaining maintenance software applications for the continuous process industries. After working with clients to help them become more proactive in their maintenance activities, he began instructing industrial plants on reliability methods and technologies to help improve the reliability of their facilities.

Over the past few years, a majority of his time has been spent developing reliability approaches with a heavy emphasis on root cause analysis (RCA). He has trained thousands of engineers and technical representatives on how to implement a successful RCA strategy at their respective facilities. He has coauthored two RCA training seminars for engineers and hourly people respectively.

Mr. Latino is also the software designer of the RCA program entitled PROACT®. PROACT® was a Gold Medal Award winner in *Plant Engineering's* 1998 and 2000 Product of the Year competitions for its first two versions on the market. He is currently working with client companies of Meridium, Inc. to integrate all of their reliability data and initiatives into an automated enterprise reliability management system.

Acknowledgments

This book would never have been possible had it not been for our father, Charles J. Latino, who had the courage to fight for Reliability Engineering early in his career with Allied Chemical in 1951 when no one would listen. He stood his ground until he proved his concepts to be of great value within the corporation. He established and directed Allied Chemical's Corporate Reliability Engineering Department in 1972. Charles retired from Allied in 1985 and purchased the center from the corporation.

Charles had the further courage to start his own Reliability Consulting firm after retirement so that he would have a business to leave his children. Having worked for Reliability Center Incorporated (RCI) for 17 years ourselves, side-by-side with our father, we could not help but become experts in the field through osmosis if through no other way. Charles has embedded tough standards and ethics into the way we conduct business and for that we are eternally thankful.

We would like to also acknowledge our mother, Marie Latino, who had the courage to support Charles in all his endeavors and endure all the business trips, missed dinners, and absence from family activities. The majority of our family now works at RCI and each child has their respective expertise to contribute. Because of our parents' understanding, we have managed to still be a cohesive family, as well as business colleagues.

We would be remiss if we did not recognize our own families for their patience and understanding in meeting publishing deadlines and doing whatever it takes to get to publication.

We would like to acknowledge the entire RCI staff for their assistance, contribution, and cooperation in making this text possible. They are a pleasure to work with and we are fortunate to have them aboard.

Finally, without our star clients who usually had to buck the system to obtain their successes, we would have nothing to tout. They are the real heroes as they put their money where their mouths were and let actions replace words. A salute to all of our clients, for their unending faith in our root cause analysis (RCA) system and our company's ability to support its implementation.

Dedications

In Loving Memory of
Joseph Raymond Latino and William T. Burns

We also dedicate this text to all of those who gave
their lives for our freedom and to those who demonstrated
their heroism during the tragic terrorist attack
on America on September 11, 2001.
May God Bless America!

Table of Contents

1 Introduction to Root Cause Analysis (RCA)

THE BUZZWORD SYNDROME

Exactly what is root cause analysis (RCA)? Is it simply another organizational buzzword that will end up being a fad, living out its half-life of six months and fading off into the sunset? We have all faced these "program-of-the-month" scenarios and find that it is difficult to implement any new thinking because of the way we have handled such new initiatives in the past. We are not only guilty of implementing such efforts ineffectively, but we are also victims to the "programs" that have been put upon us. With all of this said, we surely can understand the inherent distrust of the new corporate or facility initiative.

This is important to understand because RCA, in many instances, will be viewed as the program of the month. In this book we will show that merely having talented people performing excellent, thorough RCA's does not define success. Creating the environment for such an effort to be successful is vital to producing bottom-line results. Oftentimes we have seen management purchase expensive new equipment for testing, analysis, and inspection and demand immediate results. What they rarely consider is the environment where this technology must be applied. New technology has a tendency to scare and intimidate people. It is not accepted immediately.

When computers were first introduced, we were hesitant about switching over from the manual methods such as typing. We feared that we would lose all the information. We could see the information on the typewriter. We could not see it on the computer disk. Because we are human, we have a natural distrust of what we cannot see. In Chapter 2, we will discuss in-depth how to overcome these misconceptions about new ideas.

FOCUSING ON WHAT IS IMPORTANT

In Chapter 4 we will focus on working on the "right" event or undesirable outcome. Many RCA efforts fail because people try to be good soldiers and work on any and every failure that crops up. This is a defeatist approach, which cannot succeed. Of all the signals that reach our sensory register, we focus on a few that seem important (normal capacity is about seven "chunks" of information).* *Chunks* is the scientific term, believe it or not. So when we have 25 to 30 initiatives, responsibilities, failures, etc., it is virtually impossible to focus on a few. However, as we have all experienced,

---

* Caine, R. N., and Caine, G., (1991,1994). *Making Connections: Teaching and the Human Brain.* New York: Addison-Wesley.

TABLE 1.1
Sample Modified Failure Modes and Effects Analysis

Subsystem	Event	Mode	Frequency	Manpower	Material	Lost Profit Opportunity	Total Annual Loss
Power	Compressor failure	Trips	6/Year	$1,500	$2,500	$200	$1.224 MM
Coker	Pump failure	Bearing failure	12/Year	$700	$150	$75	$910 K

once we do have only a few items to focus on, we are typically very successful. Most RCA efforts lack focus and guidance on how to logically determine what is *most important* to the organization, vs. just *important*. Lacking focus in the beginning of such an initiative will surely doom the effort. We will present a technique called modified failure modes and effects analysis (FMEA) which assists RCA analysts in focusing on the most important issues. An example of this technique is shown in Table 1.1.

COST VERSUS VALUE

Once the organization has committed to an RCA effort, created the environment for success, and provided focusing tools for their analysts, it is time to start a comprehensive understanding of a true *"root"* cause analysis effort.

We stress *root* cause because many RCA methods exist in today's competitive environment. Many RCA methods are less expensive to your organization from an initial cost standpoint, but few are capable of providing value to your organization's bottom line.

Consider the emphasis in the late 1980s and the early 1990s and even today on cost reduction. Reengineering, right-sizing, downsizing fueled much of this; we have even heard of it referred to as *capsizing.* Such initiatives caused cost consciousness from the top to the floor. We do not think this concept is bad, but it can force rash decisions in the short term, which can be very costly in the long term.

Think about typical purchasing departments. In many instances purchasing agents are rewarded for their efforts to reduce costs. Frequently, this mentality can instill the paradigm to buy from the low cost bidder for the sake of price. This low cost product may be substandard when used in the field and have a shorter life cycle, causing premature failure and wear on other components. We have yet to run across a manufacturing operation that correlates poor performance of purchased parts in the field with purchasing practices.

When reviewing RCA methods, approaches, and software, we must ask ourselves how the utilization of this product will produce results for the organization's bottom line. The first concern should be return on investment (ROI) and the second should be initial cost. When we commission engineering projects, capital projects, or the like, the organization typically requires that a cost/benefit analysis be performed, and a quantitative ROI be determined. Why should training be any different? We have come across only a handful of companies that measure ROI for training. This is unfortunate because over $60 billion* is spent annually on occupational training and only about 20% of the training is ever implemented. This is an unbelievable waste of money.

COMPLIANCE TRAINING

We have heard horror stories of companies that train their employees on safety issues for instance. Some companies have expressed that they train their personnel on safety issues merely to comply with government regulatory agencies that require so many hours of training per employee. We call this "butt-in-seat" training! This is training merely for the sake of compliance, with no expectations of demonstrated benefits to the organization. When you talk to the employee who has sat in such classes, you hear comments like, "the mind can only take what the butt can endure," or "I've seen the same video for four years straight."

Such training is counterproductive to the culture of a progressive organization because it adds to the perception that this is just another program. Keep in mind that a program by its definition has a finite beginning and end, whereas a *process* connotes a way we do business. This is why we like to refer to RCA as a process vs. a program.

INCIDENT VERSUS PROACTIVE RCA

When getting into an RCA effort, we must consider the expected role of the analyst. Many organizations consider RCA only when an "incident" occurs. Usually an incident, or the like, is clearly defined by some regulatory agency such as OSHA, EPA, NRC, or the Joint Commission on Accreditation of Healthcare Organizations (JCAHO). Using RCA only to analyze high-visibility occurrences ends up being a reactive use of a proactive tool. The only reason this particular incident will get attention is because the government, the lawyers, and the insurance companies may be involved. In cases such as these, an analysis will be performed whether we like it or not.

What about all those chronic issues that occur in our organizations daily, and sometimes even on each shift? Usually no one is hurt and there is minimal, if any, equipment or physical damage. Typically, the result is a short lapse in production or process time, then things are started up again and everything and everyone is happy. Because the labor, equipment, and production losses are perceived as relatively small, we tend to accept these chronic events as a cost of doing business. We call this "retiring the chronic failures to the pasture of the budget."

Think about our budgets. We almost always have a "fat" category, and it is usually marked as an "R" for routine or an "M" for miscellaneous. We actually

* ASTD Web Site, Training Statistics, 1996.

budget for such chronic failures or events to occur and plan our production strategies around them. In effect, we will make our goals in spite of these nuisances. If we were to add up these nuisances and hang a price tag on them on an annual basis, we would find that they are more costly than any sporadic or one-time event. We will discuss in this book that we feel many corporations today may be focusing on the wrong events. This is a huge waste of resources that do not generate the necessary numbers on the bottom line. Focusing on not accepting these chronic events and encouraging the analysts to eliminate their recurrence is a proactive use of RCA.

THE PROACT® METHOD

In Chapter 5 we will begin to explore in-depth the disciplined thought process involved in conducting a true root cause analysis (RCA). PROACT® is an acronym that stands for the following:

- **PR**eserving Event Data
- **O**rdering the Analysis Team
- **A**nalyzing the Data
- **C**ommunicating Findings and Recommendations
- **T**racking for Results

We have used this process to assist corporations around the world in saving hundreds of millions of dollars on the bottom line. The process is simple and to the point; the key to its success is the adherence to the discipline of following the rules.

PRESERVING EVENT DATA

In any investigative or analytical occupation, whether an NTSB investigator, a police detective, a scientist, or a doctor, the above steps must be performed in order to be successful. Think about it — we cannot begin an analysis without any data. Therefore, a disciplined effort must be in place to identify what data is necessary, how to collect it, and when to collect it by. Only then can you begin an analysis.

In Chapter 5, we will explore what we call the 5-Ps (parts, position, paper, people, and paradigms) of data collection. Any necessary information from an undesirable outcome can be categorized in some manner under one of these headings.

ORDERING THE ANALYSIS TEAM

Next, we must build a team to review the data. The team must consist of a lead analyst or facilitator and various experts and nonexperts as core team members. The diversity of the team is very important and will be discussed at length in Chapter 6. An effective and productive team shall also have structure and direction. This means that rules will exist such as team rules of conduct, a team charter or mission, and critical success factors (CSF) that delineate when we will know that the team has been successful and completed its mission. These issues will also be discussed in detail in later chapters.

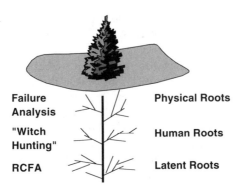

FIGURE 1.1 Levels of root cause.

ANALYZING THE DATA

Once we have collected the necessary data to begin an analysis and assembled the ideal team, it is time to take the pieces of the puzzle (the data) and put it all together so that it makes logical sense. This is where the analytical process takes place.

Have you ever been involved in a situation where everything seemed chaotic? Whether it was an explosion or fire, a "code" in the hospital, a failure of a piece of equipment, everything seems chaotic because we do not know what happened. There are usually more variables than we know what to do with, and it seems impossible that we will ever discover why the undesirable outcome occurred.

This is what RCA is all about, making order out of chaos. We will go into research that proves that everything does happen for a reason and usually many errors occur before an undesirable outcome occurs. Through the use of a "logic tree," we can deductively trace backwards from known facts to *how* such occurrences could have happened. Then we can use our pieces of the puzzle to prove and disprove what actually *did* happen.

THE ROOT CAUSES

This reiterative process of developing hypotheses and proving or disproving them is based on experimentation and is a vital part of any investigative or analytical process. Ultimately, the process will conclude with the identification of physical, human, and latent root causes (see Figure 1.1).

Physical root causes are the tangibles that can be seen and usually are the stopping point of most organizations that say they do RCA. Human roots are involved in virtually every undesirable outcome that occurs in our environment. We say virtually because acts of God could be an exception (although there are circumstances where we can better protect ourselves from acts of God such as by installing lighting rods). Most undesirable outcomes will be the result of human errors of commission or omission. This simply means that either we made an inappropriate decision or we overlooked the need to make a decision. Such instances are repeated every day as a part of being human and will be elaborated on in Chapter 8. Stopping and identifying a human root cause will almost always result in the perception of "witch hunting" by the individual or group pointed out, and is most certainly NOT recommended. That would make any RCA effort counterproductive and immediately create a general resistance to the overall effort.

Finally, we must face the fact that we should not be as interested in WHO made an error in decision-making as in WHY they made that decision. It has been our experience that most people who go to work every day and draw their paychecks to support their livelihood do not intend to cause undesirable outcomes to occur. We are of the belief that all people are good and well intentioned, until proven otherwise. Most people who have made bad decisions intended in their logic to make the right decision. We must delve deeper into what information was used to arrive at a poor decision, because this is the true root cause, or latent root. Every organization has management systems or organizational systems. These are the policies, procedures, training systems, purchasing systems, etc. that are put in place to help its employees make better decisions. However, we are often not disciplined enough to be able to keep such systems up to date. Do we ever see where new technology was introduced, yet the appropriate procedures were not changed to accommodate that technology? Under this scenario we would be adhering to obsolete procedures that were incorrect for the circumstance, and ultimately an undesirable event would result. We will discuss the concept of latency in detail in Chapter 8.

COMMUNICATING FINDINGS AND RECOMMENDATIONS

We call the point between obtaining solid causes and getting something done about them the halfway point. We believe that just as much headache and effort are put into identifying real root causes as in trying to get legitimate recommendations implemented in the field. Many organizations have such bureaucracies in place that trying to get proactive recommendations implemented is a nightmare. Such environments discourage innovation and encourage frustration in attempts to do the right thing. In Chapter 9 we will discuss at length how to identify and overcome these barriers.

TRACKING FOR RESULTS

The goal of any investigation or analysis should not be merely to identify what went wrong, but to ensure that the risk of recurrence is eliminated by the implementation of recommendations to correct identified system deficiencies. Only when the identified metrics have improved should the effort be deemed successful. See Figure 1.2. This task requires the identification of metrics to monitor bottom-line performance. Such metrics will be identified in Chapter 10.

AUTOMATING RCA

The PROACT® approach is very disciplined and therefore requires administrative skills such as organization. Analysis cycle times should be a concern to any analyst. When administrative tasks burden an analyst to the point that it drags the analysis on and on, consideration should be given to identifying automation sources or software that can alleviate paperwork and expedite the completion of an analysis.

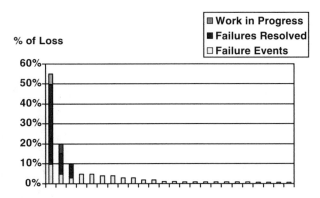

FIGURE 1.2 Sample tracking mechanism using modified FMEA.

Think of how data-intensive a detective's work is in building a solid case for court. Then think of the scrutiny that the data goes through in the courtroom. Good organizational skills and proper documentation habits can be the difference between a successful RCA and one that does not produce results.

In Chapter 11 we will outline what key characteristics and features such automation should possess and how they benefit the analysis team and the analysis itself.

RCA KNOWLEDGE MANAGEMENT

Think about what we normally do in our everyday work environment where RCA is concerned. Normally, at a minimum, we document our analysis for reporting purposes to our upper management. There may even be a limited distribution to copyholders. But ultimately, the report ends up in a filing cabinet rarely to be seen again.

This is where we must acknowledge what our RCA goals are. From a knowledge management perspective, we must seek to optimize the use of the successful logic developed to solve a problem. After all, a successful RCA involves the input of many experienced and intelligent people. For the results of their time together to gather dust in a filing cabinet is a waste.

So now we must face a decision about how we define the success of our RCA effort. Are we content with merely eliminating the risk of recurrence of a single event, or do we wish to transfer the results of the successful RCA (knowledge) to others in the corporation who may benefit from the information in their facilities? The latter is the more desirable from a corporate standpoint. In this manner we treat RCA as an approach rather than a project. Figure 1.3 demonstrates how electronically storing the successful logic allows such information to be stored efficiently.

CASE HISTORIES

In conclusion, we will present six case histories of where PROACT® has been applied per the written text and produced quantum, bottom-line results in a very short time

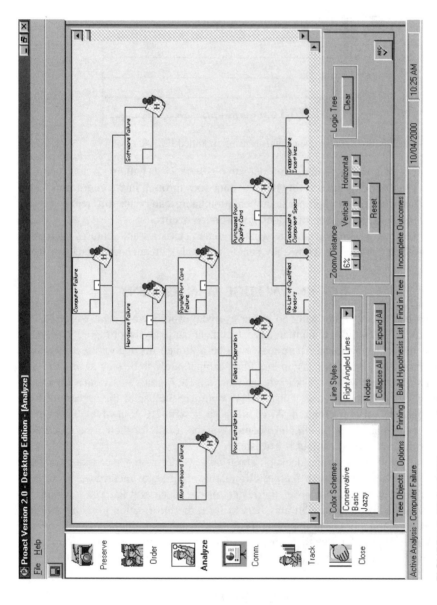

FIGURE 1.3 PROACT™ screen shot.

frame. Finally, in Chapter 12, we will identify the driving force behind these analyses and analysts that made them successful.

This book is written based on experience in implementing the PROACT® method. We do not present ourselves as strictly educators, but also as practitioners of and believers in a field-proven method that does indeed drive an organization towards, *"eliminating the need to do reactive work!."*

2 Creating the Environment for RCA to Succeed: The Reliability Performance Process (TRPP)©

TRPP©* is a training model developed by Reliability Center, Inc. (RCI). It encompasses not only the elements about specific training objectives necessary to be successful, but it also outlines the specific requirements of the executives/management, the *champions* and the *drivers* who are accountable for creating the environment for RCA to be successful.

We will be outlining specific information from TRPP© that is pertinent to creating the environment for RCA to succeed.

THE ROLE OF EXECUTIVE MANAGEMENT IN RCA

For the implementation of any initiative in an organization, the path of least resistance is typically from the top down rather than from the bottom up. The one thing we should always be cognizant of is the fact that no matter what the new initiative is, it will likely be viewed by the end user as the program of the month. This should always be in the back of our minds when developing implementation strategies.

Our experience is that the closer we get to the field where the work is actually performed, the more skeptics we will encounter. Every year new organizational "buzz" fads emerge, and the executives hear and read about them in trade journals, magazines, and business texts. Eventually directives are given to implement these fads, and by the time they reach the field, the well-intentioned objectives of the initiatives are so diluted and saturated from miscommunication that they are viewed as non-value-added work and a burden to an existing workload. This is the misconception of the end user that must be overcome to be successful at implementing RCA.

Oftentimes when we look at instituting these types of initiatives, we look at them strictly from the shareholders' view and work backwards. Do not get us wrong, we are not against new initiatives that are designed to change behavior for the betterment of the corporation. This process is necessary to progress as a society. However, the manner in which we try to attain the end is what has been typically ineffective.

In the eyes of the field or end user, we must distinguish this initiative from those that have failed. We must look at the reality of the environment of the people who

---

* Reliability Center, Inc. "The Reliability Performance Process." Hopewell: RCI, 1997.

will make the change happen. How can we change the behavior of a given population to reflect those behaviors that are necessary to meet our objectives?

Let's take an example. If I am a maintenance person in an organization and have been so for my entire career, I am expected to repair equipment so that we can make more products. As a matter of fact, my performance is measured by how well I can make the repair in the shortest time frame possible. I am given recognition when emergencies occur and I respond almost heroically.

Now comes along this root cause analysis (RCA) initiative and they want me to participate in making sure that failures do not occur anymore. In my mind, if this objective is accomplished, I am out of a job! Rather than be perceived as NOT being a team player, I will superficially participate until the program of the month has lived out its six-month life and then go on with business as usual. We have seen this scenario repeatedly and it is a very valid concern based on the reality of the end user. This perception must be overcome prior to implementing an RCA initiative.

Let's face the fact that we are in a global environment today. We must compete not only domestically, but now with foreign markets. Oftentimes these markets have an edge in that their costs to produce are significantly less than here in the U.S. Maintenance, in its true state, is a necessary evil to a corporation. But when equipment fails, it generally holds up production, which holds up delivery, which holds up profitability. Imagine a world where the only failures that occurred were wear-out failures that were predictable. This is a world that we are moving towards, as precision environments become more the expectation. As we move in this direction, there will be less need for maintenance type skills on a routine basis. What about the area of reliability engineering? Most organizations we deal with never have the resources to properly staff their reliability engineering groups. There are plenty of available roles in the field of reliability. Think about how many reliability jobs are available: vibration analysts, failure analysts, infrared thermographers, metallurgists, designers, inspectors, nondestructive testing specialists, and many more.

We are continually intrigued by the most frequently used objection to using RCA in the field from our students, "I don't have time to do RCA." If you think hard about this statement, it really is an oxymoron. Why do people typically not have time to do RCA? They are so busy fighting fires, they do not have time to analyze why the event occurred in the first place. If this remains as a maintenance strategy, then the organization will never progress, because no level of dedication is being put towards getting rid of the need to do the reactive work!

So how can executives get these very same people to willingly participate in a new RCA initiative?

1. It must start with an executive putting a rubber stamp on the RCA effort and outlining specifically what the expectations are for the process and a time line for when bottom-line results are expected.

2. The approving executive(s) should be educated in the RCA process themselves, even if it is an overview version. Such demonstrations of support are worth their weight in gold because the users can be assured that the executives have learned what they are learning and agree and support the process.

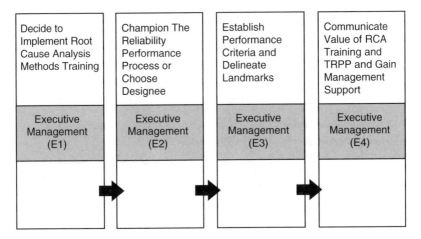

FIGURE 2.1 TRPP© executive management roles.

3. The executive responsible for the success of the effort should designate a champion of the RCA effort. This individual's roles will be outlined later in this chapter.
4. It should be clearly delineated how this RCA will benefit the company, but more important, it should also delineate how it will benefit the work life of every employee.
5. Next the executive should outline how the RCA process will be implemented to accomplish the objectives and how management will support those actions.
6. A policy or procedure should be developed to institutionalize the RCA process. This is another physical demonstration of support that also provides continuity of the RCA application and perceived staying power. It gives the effort perceived staying power because even if there is a turnover in management, institutionalized processes have a greater chance of weathering the storm.
7. However, the most important action an executive can take to demonstrate support is to sign a check. We believe this is a universal sign of support.

See Figure 2.1 for an illustration of TRPP executive management roles.

THE ROLE OF AN RCA CHAMPION

Even if all the above actions take place, success is not automatically ensured. How many times have we all seen a well-intentioned effort from the top try to make its way to the field and fail miserably? Typically, somewhere in the middle of the organization the translation of the original message begins to deviate from its intended path. Miscommunication of the original message is a common reason why some very good efforts fail!

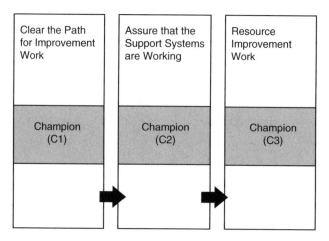

FIGURE 2.2 TRPP© champion roles.

If we are proactive in our thinking and we foresee such a barrier to success, then we can plan for its occurrence and avoid it. This is where the role of the RCA champion comes into play.

There are three major roles of an RCA champion:

1. The champion must administer and support the RCA effort from a management standpoint. This includes ensuring that the message from the top to the floor is communicated properly and effectively. Any deviations from the plan will be the responsibility of the champion to align or get back on track. This person is truly the "champion" of the RCA effort.
2. The second primary role of the RCA champion is to be a mentor to the drivers and the analysts. This means that the champion must be educated in the RCA process and have a thorough understanding of what is necessary for success.
3. The third primary role of the RCA champion is to be a protector of those who use the process and uncover causes that may be politically sensitive. Sometimes we refer to this role as providing "air cover" for ground troops. In order to fulfill this responsibility, the RCA champion must be in a position of authority to take a defense position and protect the person who uncovered these facts.

These roles are illustrated in Figure 2.2.

Ideally, this would be a full time position. However, in reality, we find it typically to be a part-time effort for an individual. In either situation we have seen it work; the key is that it must be made a priority to the organization. This is generally accomplished if the executives perform their designed tasks set out above. Actions do speak louder than words. When new initiatives come down the pike and the workforce sees no support, then it becomes another "they are not going to walk-the-talk" issue. These are viewed as lip service programs that will pass over time. If the

RCA effort is going to succeed, it must first break down the paradigms that currently exist. It must be viewed as different than the other programs. This is also the RCA champion's role in projecting an image that this is different and will work.

The RCA champion's additional responsibilities include ensuring that the following responsibilities are carried out:

1. Selecting and training RCA drivers who will lead RCA teams. What are the personal characteristics that are required to make this a success? What kind of training do they need to provide them the tools to do the job right?
2. Developing management support systems such as:
   a. RCA performance criteria — What financial returns does the corporation expect? What are the time frames? What are the landmarks?
   b. Providing time — In an era of reengineering and lean manufacturing, how are we going to mandate that designated employees WILL spend 10% of their week on RCA teams?
   c. Process the recommendations — How are recommendations from RCAs going to be handled in the current work order system? How does improvement (proactive) work get executed in a reactive work order system?
   d. Provide technical resources — What technical resources are going to be made available to the analysts to prove and disprove their hypotheses using the "whatever it takes" mentality?
   e. Provide skill-based training — How will we educate RCA team members and ensure that they are competent enough to participate on such a team?
3. The champion shall also be responsible for setting performance expectations. The champion should draft a letter that will be forwarded to all students that attend the RCA training. The letter should clearly outline exactly what is expected of them and how the follow-up system will be implemented.
4. The champion should ensure that all training classes are kicked off whether by themselves, an executive, or other person of authority, giving credibility and priority to the effort.
5. The champion should also be responsible for developing and setting up a recognition system for RCA successes. Recognition can range from a letter from an executive to tickets to a ball game. Whatever the incentive is, it should be of value to the recipient.

Figure 2.3 illustrates these responsibilities.

Needless to say, the role of a champion is very critical to the RCA process. The lack of a champion is usually why most formal RCA efforts fail. There is no one leading the cause or carrying the RCA flag. Make no bones about it, if an organization has never had a formal RCA effort, or had one that failed, such an endeavor is an uphill battle. Without an RCA champion, sometimes you can feel like you are on an island by yourself.

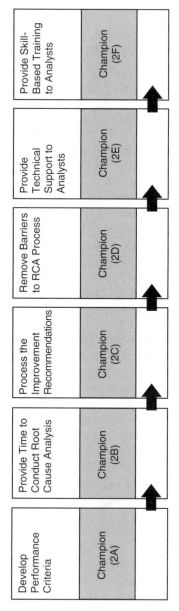

FIGURE 2.3 TRPP© additional roles of champions/management.

THE ROLE OF THE RCA DRIVER

RCA drivers can be synonymous with the RCA team leader. These are the people who organize all the details and are closest to the work. Drivers carry the burden of producing bottom-line results for the RCA effort. Their teams will meet, analyze, hypothesize, verify, and draw factual conclusions as to why undesirable outcomes occur. Then they will develop recommendations or countermeasures to eliminate the risk of recurrence of the event.

All the executive, manager, and champion efforts to support RCA are directed at supporting the driver's role to ensure success. The driver is in the unique position of dealing directly with the field experts, the people that will make up the core team. The personality traits that are most effective in this role as well as a core team member role will be discussed at length in Chapter 7.

From a functional standpoint the RCA driver's roles are:

1. Making arrangements for RCA training for team leaders and team members — This includes setting up meeting times, approving training objectives, and providing adequate training rooms.
2. Reiterating expectations to students — Clarify to students what is expected of them, when it is expected, and how it will be obtained. The driver should occasionally set and hold RCA class reunions. This reunion should be announced at the initial training so as to set an expectation of demonstrable performance by that time.
3. Ensuring that RCA support systems are working — Notify the RCA champion of any deficiencies in support systems and see that they are corrected.
4. Facilitating RCA teams — The driver shall lead the RCA teams and be responsible and accountable for the team's performance. The driver will be responsible for properly documenting every phase of the analysis.
5. Documenting performance — The driver will be responsible for developing the appropriate metrics to measure performance against. This performance shall always be converted from units to dollars when demonstrating savings, hence success.
6. Communicating performance — The driver shall be the chief spokesperson for the team, presenting updates to management as well as to other individuals on-site and at other similar operations that could benefit from the information. The driver shall develop proper information distribution routes so that the RCA results get to others in the organization that have had similar occurrences.

See Figure 2.4 for a graphic presentation of these roles.

The driver is the last of the support mechanisms that should be in place to support such an RCA effort. Most RCA efforts that we have encountered are put together at the last minute as a result of an incident that just occurred. We discussed this topic earlier regarding using RCA as only a reactive tool.

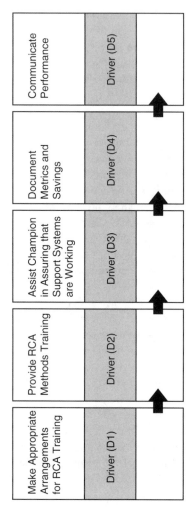

FIGURE 2.4 TRPP© driver roles.

A structured RCA effort should be properly placed in an organizational chart. Because RCA is intended to be a proactive task, it should reside under the control of a structured reliability department (see Figure 2.5). In the absence of such a department, it should report to a staff position such as a VP of operations or VP of engineering. Whatever the case may be, ensure that an RCA effort is never placed under the control of a maintenance department. By its nature, a maintenance department is a reactive entity. Its role is to respond to the day-to-day activities in the field. The role of a true reliability department is to look at tomorrow, not today. Any proactive task assigned to a maintenance department is typically doomed from the start.

This is the reason that when "reliability" became the buzzword of the mid-1990s that many maintenance engineering departments were renamed reliability departments. The same people resided in the department and they were performing the same jobs; their title was changed but not their function. If you are an individual who is charged with the responsibility of responding to daily problems and also seizing future opportunities, you are likely to never get to realize those opportunities. Reaction wins every time in this scenario.

Now let us assume that at this point we have developed all the necessary systems and personnel to support an RCA effort. How do we know what opportunities to work on first? Working on the wrong events can be counterproductive and yield poor results. In the next chapter we will discuss a technique for selling your reasons for working on one event vs. another.

SETTING FINANCIAL EXPECTATIONS:
THE REALITY OF THE RETURN

As discussed earlier, one of the roles of the champion is to delineate financial expectations of the RCA effort. This will obviously vary with the key performance indicators (KPI) of each firm, but in this section we will look at providing a typical business case to justify implementing an RCA effort.

Because the costs to implement such an effort will vary based on each facility, their product sales margin, their labor costs, and the training costs (in-house vs. contract), we will base our justifications on the following assumptions:

1. Employee cost
   a. Loaded cost of hourly employee: $50,000/year
   b. Hourly employees will spend 10% of their time on RCA teams
   c. Loaded cost of full-time RCA driver (salaried): $70,000/year
   d. RCA driver will be a full-time position
   e. RCA training costs (hourly): $400/person/day
   f. RCA training costs (salaried): $500/person/day
   g. Population trained: Per 100 trained
2. RCA return expectations
   a. Train 100 hourly employees in RCA methods
   b. Train one salaried employee to lead RCA effort

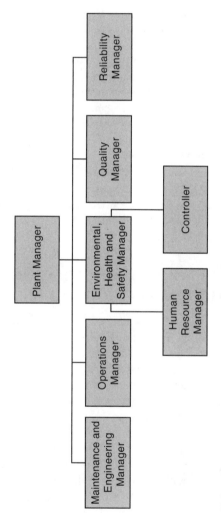

FIGURE 2.5 Ideal position for reliability in plant.

    c. Critical mass (assumption): 30% of those trained will actually use the RCA method in the field. This results in 30 personnel trained in RCA methods actually applying the method in the field (100 trained × 30% applying).

    d. Let's assume the 30 personnel applying the RCA method are working in teams of three at a minimum. This results in 10 RCA teams applying the methodology in the field (30 personnel/3 per team).

    e. Each RCA team will complete one analysis every two months. This results in 60 completed analyses per year (10 RCA teams × 6 analyses/year).

    f. Each "significant few" (to be discussed in Chapter 3) analysis will net a minimum of $50,000 annually. This results in an annual return of $3,000,000 per 100 people trained in RCA methods.

3. The costs of implementing RCA

Year 1:

a. Training 100 hourly employees in three days of RCA:	$120,000
b. Training one salaried person in five days of RCA:	$2,500
c. 10% of 30 hourly employees' time per week, annually:	$150,000
d. Salary of RCA driver/year:	$70,000
e. Total RCA implementation costs for year 1:	$342,500

Year 2:

a. Training 100 hourly employees in three days of RCA:	$0
b. Training one salaried person in five days of RCA:	$0
c. 10% of 30 hourly employees' time per week, annually:	$150,000
d. Salary of RCA driver/year:	$70,000
e. Total RCA implementation costs for year 2:	$220,000

    All costs of resources to prove the hypotheses and implement recommendations are considered sunk costs. Technical resources are currently available and budgeted for, regardless of RCA. Also, recommendations from RCA generally result in the implementation of organizational system corrections. For instance rewriting procedures, providing training, upgrading testing tools, restructuring incentives, etc. These types of recommendations are not generally considered as capital costs. Capital costs resulting from RCA, in our experience, are not the norm, but the exception.

4. Return-on-investment

a. Total expected return — Year 1:	$1,500,000
b. Total expected costs — Year 1:	$342,500
c. ROI Year 1:	437%

The expected return for Year 1 is based on the assumption that it will take six months to train all involved and get up to speed with actually implementing RCA and the associated recommendations. This is the reasoning for cutting this expectation in half for the first year.

d. Total expected return — Year 2:	$3,000,000
e. Total expected costs — Year 2:	$220,000
f. ROI Year 2:	1360%

As we can tell from these numbers, the opportunities are left to the imagination. They are real, they are phenomenal to the point that they are unbelievable. To review the process we just went through, look at the conservativeness built in:

1. Only 30% of those trained will actually apply the RCA method.
2. Students will only spend 10% of their time on RCA.
3. Students will work in teams of three or more.
4. Students will only complete one RCA every two months.
5. Each event will only net $50,000/year.

Use this same cost-benefit thought process and plug in your own numbers to see if the ROIs are any less impressive. Using the most conservative stance, it would appear irrational not to perform RCA in the field. How many of our engineering projects would be turned down if we demonstrated to management an ROI ranging from 437% to 1360%? Not many!

3 Failure Classification

PROBLEMS VERSUS OPPORTUNITIES

To begin discussing the issue of root cause analysis, we must first set the foundation with some key terminology. Let's begin by discussing the key differences between problems and opportunities. There are many people who tend to use these terms interchangeably. The truth is these terms are really at opposite ends of the spectrum in their definition.

A problem can be defined as a negative deviation from a performance norm. What exactly does this mean? It simply means that we cannot perform up to the normal level or standard that we are used to. For example, let's assume we have a widget factory. We are able to produce 1000 widgets per day in our factory. At some point we experience an event that interrupts our ability to make widgets at this level. This means that we have experienced a negative deviation from our performance norm, which in this case is 1000 widgets. See Figure 3.1.

An opportunity is really just the opposite of a problem. It can be defined as a chance to achieve a goal or an ideal state. This means that we are going to make some changes to increase our performance norm or status quo. Let's look back at our widget example. If our normal output was 1000 widgets per day, then any changes we make to increase our throughput would be considered an opportunity. So if we eliminate certain bottlenecks from the system and start to produce 1100 widgets in a day, this would be considered an opportunity (see Figure 3.2).

Now let's put these terms into perspective. When a problem occurs and we take action to fix it, do we actually improve or progress? The answer to this question is an emphatic *no*. When we work on problems we are essentially working to maintain the status quo or performance norm. This is synonymous with the term *reaction*. We react when a problem occurs to get things back to their normal state. If all we do is work on problems, we will never be able to progress. In our dealings with companies all over the world, we often ask the question, "How much time do you spend being reactive rather than proactive?" Just about everyone surveyed will answer 80% reactive and 20% proactive. If this is true, then there is very little progress being made. This would seem to be a key indicator as to why most productivity increases are minimal from year to year.

Let's consider opportunities for a moment. When we work on opportunities do we progress? The answer is *yes*! When we achieve opportunities we are striving to raise the status quo to a higher level. Therefore, to progress we have to begin taking advantage of the numerous opportunities presented to us. So if working on problems is reactive, then working on opportunities is proactive. See Figure 3.3.

So the answer is simple. We should all start working on opportunities and disregard problems, right? Why can't we do this? There are many reasons but a few are obvious. We are more aware of problems since they take us away from our

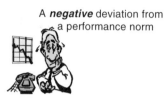

A **negative** deviation from
a performance norm

FIGURE 3.1 Problem definition graph.

A chance to achieve a
goal or an ideal state.

FIGURE 3.2 Opportunity definition graph.

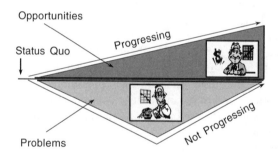

FIGURE 3.3 Opportunity graph.

normal operation. Therefore we give them more attention and priority. We can always put an opportunity off until tomorrow, but problems have to be addressed today. There is also the issue of rewards. People who are good reactors that come in and save the day tend to get pats on the back and the old "atta-boys." What a great thing from the reactor's perspective — recognition, overtime pay, and most important job security. We have seen many cases where the person who tries to prevent a problem or event from occurring gets the cold shoulder while the person who comes in after the event has occurred gets treated like a king. Not to say we should not reward good reactors, but we also have to reinforce good proactive behavior as well.

Then there is the risk factor. Which are more risky, problems or opportunities? Opportunities are always more risky because there are many unknowns. With problems there are virtually no unknowns. We usually have fixed the problems before, so we certainly have the confidence to fix it again. I once had a colleague who said, "… when you get really good at FIXING something, you are getting way too much practice." In a perfect world, we should have to pull the manual out to see what steps to take to fix the problem. How many times do we a see a craftsman, or even a doctor for that matter, pulling out the manual to troubleshoot a problem? People today do not want to take a lot of chances with their careers so opportunities begin to look like what we call "career-limiting" activities.

So with that said, we have to figure out a way of changing the concept that reactive is always more important then proactive work. This means that opportunities are just as important as, if not more important than, problems.

SPORADIC VERSUS CHRONIC EVENTS

Let's switch gears and talk about the different types of failures or events that can occur. Incidentally, when we talk about failures we are not always talking about machines or equipment. Failure can also be process upsets, administrative delays, quality defects, or even customer complaints. There are two basic categories of failures that can exist, sporadic and chronic. Let's look at each of these categories is greater detail.

A sporadic occurrence usually indicates that a dramatic event has occurred. For example, maybe we had a fire or an explosion in our manufacturing plant or we just lost a long-standing contract to a competitor. These events tend to demand a lot of attention. Not just attention, but urgent and immediate attention. In other words, everyone in the organization knows that something bad has happened. The key characteristic of sporadic events is that they happen only one time with usually one mode. There is one mechanism at work that has caused this event to occur. This is very important to remember. Sporadic failures have a very dramatic impact when they occur, which is why many people tend to apply financial figures to them. For instance, you might hear someone say, "We had a $10,000,000 failure last year."

Sporadic events are very important and they certainly do cost a lot of money when they occur. The reality, however, is that they do not happen very often. If we had a lot of sporadic events, we certainly would not be in business very long. Sporadic losses can also be distributed over many years. For example, if the engine in your car fails and you need to replace it, it will be a very costly expense, but you can amortize that cost over the remaining life of the car.

Chronic events on the other hand are not very dramatic when they occur. These types of events happen over and over again. They happen so often that they actually become a cost of doing business. We become so proficient at working on these events that they actually become part of the status quo. We can produce our normal output in spite of these events.

Let's look at some of the characteristics of chronic events. Chronic events are accepted as part of the routine. We accept the fact that they are going to happen. In a manufacturing plant, we will even account for these events by developing a maintenance budget. A maintenance budget is in place to make sure that when routine events occur we have money on hand to fix them. These events do demand attention but usually not the attention a big sporadic event would. The key characteristic of a chronic event is the frequency factor. These chronic events happen over and over again for the same reason or mode. For instance, on a given pump, the bearing may fail three or four times a year or if you a have a bottle filling line and the bottles

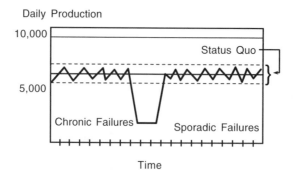

FIGURE 3.4 The linkage.

continually jam, that would be considered a chronic event. Chronic events tend not to get the attention of sporadic events because as individual occurrences, they are usually not very costly. Therefore, rarely would we ever assign a dollar figure to an individual chronic event.

What most people fail to realize is the tremendous effect the frequency factor has on the cost of chronic failures. A stoppage on a bottling line due to a bottle jam may only take 5 minutes to correct when it occurs. If it happens 5 times a day, then we are looking at 152 hours of downtime per year. If an hour of downtime costs $10,000, then we are looking at a cost of approximately $1,520,000. As we can see, the frequency factor is very powerful. But since we tend to only see chronic events in their individual state, we sometimes overlook the accumulated cost. Just imagine if we were to go into a facility and aggregate all of the chronic events over a year's time and multiply their effects by the number of occurrences. The yearly losses would be staggering.

Let's take a look at how chronic and sporadic events relate to the discussion on problems and opportunities. Sporadic events by their definition take us below the status quo and tend to take an extended period of time to restore. When we restore we get back to the status quo. This is very much like what happens when we react to a problem. The problem occurs and we take some action to get back to the status quo. Chronic events, on the other hand, happen so routinely that they actually become part of the status quo. Therefore, when they occur, they do not take us below our performance norm. If, in turn, we were to eliminate the chronic or repetitive events, then the elimination would actually cause the status quo to improve. This improvement is the equivalent of realizing an opportunity. So by focusing on chronic events and eliminating the causes and not simply fixing the symptoms, we are really working on opportunities. As we said before, when we work on opportunities, the organization progresses.

Now that we know that eliminating chronic events can cause the organization to progress, we have to look at the significance of chronic events. Sporadic events by their very nature are high-profile and high-cost events. But we can amortize that cost over a long period of time so that the effect is not as severe. Consider that the engine in your car blew up and you had to replace it. To the average motorist, this would be a sporadic event. But if we amortize the cost over the remaining life of the car it becomes less of a burden. Chronic events on the other hand have a relatively low impact on their individual basis but we often overlook their true impact. If we were to aggregate all of the chronic events from a particular facility and look at their total cost over a one-year period, we would see that their impact is far more significant than any given sporadic event, simply due to the frequency factor.

Let's consider how all of the events actually affect the profitability of a given facility. As we all know, we are all in business to make a profit. When a sporadic event occurs, it actually affects the profitability of a facility significantly the year that it occurs, but once the problem has been resolved, profitability gets back to normal. The dilemma with chronic events is that they usually never get resolved so they affect profitability year after year. If we were to eliminate such events instead of just reacting to their symptoms, we could make great strides in profitability. Imagine if we had ten facilities and we were able to reduce the number of loss events in order to obtain 10% more throughput from each of those facilities. In essence we would have the capacity of one new facility without spending the capital dollars. That is the power of resolving chronic issues.

Let us give an example of a chronic event success story. In a large mining operation, the management wanted to uncover their most significant chronic events. In this operation, they have a large crane or "drag line" as they call it. This drag line mines the surface for product. The product is then placed on large piles where a machine called a *bucket wheel* moves up and down the pile putting the product onto a conveyor system. This is where the product is taken downstream to another process of the operation. One day, one of the analysts was talking to one of the field maintenance representatives who said they spend a majority of time resetting conveyor systems whose safety trip cord was pulled. They estimated that this activity took anywhere from 10 to 15 minutes to resolve. Now this individual did not see this activity as a "failure" by any means. It was just part of the job he had to do. Upon further investigation, it was discovered that other people were also resetting tripped conveyors. By their estimates, this was happening approximately 500 times a week to the tune of about $7,000,000 per year in lost production. Just by identifying this as an undesirable event allowed them to take instant corrective action. By adding a simple procedure of removing large rocks with a bulldozer prior to bucket wheel activity, approximately 60% of the problem went away. These types of stories are not uncommon. We get so caught up in what we are doing that we sometimes miss the things that are so obvious to outsiders.

We were working with a major oil company that was trying to reduce its maintenance budget. So they hired our firm to teach them the methods being explained in this text. This manager opened the three-day session by stating that he had been mandated by his superiors to reduce the maintenance budget significantly. He told them that the maintenance budget for this particular facility was approximately $250,000,000. He went on to explain that some analysis was done of the budget to find out how the money was being spent. It turns out that 85% of the money was spent in increments of $5,000 or less. By his estimate he was spending about $212,000,000 in chronic maintenance losses. This was just maintenance cost, not lost production cost!

TABLE 3.1
Options to Reduce Maintenance Budget

Scenerio: Oil Refinery Example

Annual maintenance cost	$250,000,000	
Chronic losses	85.00%	Increments of $5,000 or less
Total	$212,500,000	

Reduce the need for the work	Option A	
20.00%	$42,500,000	Net Savings
30.00%	$63,750,000	Net Savings
40.00%	$85,000,000	Net Savings

Reduce people	Option B
Employees	1,500
Average loaded salary	$75,000
Reduce employees by 7%	105
Net savings	$7,875,000

So he told the 25 engineers in the training class that he had two options to reduce this maintenance cost:

1. He could eliminate the need to do the work in the first place.
2. He could just eliminate maintenance jobs.

He said that if they could eliminate the need to do the work in the first place (e.g., reduce the amount of chronic or repetitive failures) then he felt that they could reduce the maintenance expenditures by about 20%. This would be a savings of about $42,000,000. If they were really successful, they could eliminate 30%, or $63,000,000.

He went on to say that if he took option 2 and let approximately 100 maintenance people go, that would probably net the company about $7,500,000, but he would have the same, if not more, work and fewer people to address the additional work. See Table 3.1 for an illustration of these options. To make a long story short, the people in the training class opted for option 1, reducing the need to do the work by using their abilities to solve problems!

So to sum up this discussion on failure classification, let's look at the key ideas presented. We live in a world of problems and opportunities. We would all love to take advantage of every opportunity that came about, but it seems as if there are too many problems confronting us to take advantage of the opportunities. A good way to take advantage in a business situation is to eliminate the chronic or repetitive events that confront us each and every day. By eliminating this expensive, non-value-added work, we are really achieving opportunities as well as adding additional time to eliminate more problems. In the next chapter we will discuss a method for uncovering all of the events for a given process and delineating which of those events are the most significant from a business perspective.

RCA AS AN APPROACH

As we mentioned briefly in the Introduction, RCA is certainly applicable to both chronic and sporadic events in any industry. However, focusing on RCA as only an incident or accident tool is not optimizing its potential for the corporation. Using RCA in this fashion limits its effectiveness and treats it as an off-the-shelf tool for reactive situations.

When using RCA as an approach, we seek to break the paradigm that chronic events are an accepted cost of doing business because they are compensated for in the budget. We seek to solve these chronic events down to their root causes and pass the knowledge on to others in the corporation who may be accepting them as a cost of doing business as well. This is the knowledge management and transfer component of RCA that we discussed earlier.

4 Failure Modes and Effects Analysis (FMEA): The Modified Approach

With all the noise and distraction of a reactive work environment, it is sometimes easy to overlook the obvious. For instance, if we wanted to perform a root cause analysis (RCA) on an event, would we know which event was the most significant or costly? Experience demonstrates that we would not. In a reactive environment, we naturally become focused on the short term. We tend to look at the problems or events that just happened and naturally think they are the most significant. This is a problem because what happened yesterday, in most cases, is not the most significant or compelling issue. We need to take a more macro look at the situation.

This is where we need a technique to help us take a global look at our situation and assess *all* of the events and assess their individual impact on performance. There is a technique that will help us do just that. It is called *failure modes and effects analysis,* or FMEA as it is known in the industry. This technique was developed in the aerospace industry to determine what failure events could occur within a given system (e.g., a new aircraft) and what the associated effects would be if it did occur. This technique, albeit effective, is very man-hour intensive. It is estimated that a typical FMEA in the aerospace industry takes anywhere from 50 to 100 man-years to perform. There are many good reasons that this technique takes so long to complete and there are significant benefits to performing one in this industry. This technique is far too laborious to be performed in most industries such as the process and discrete manufacturing arenas. Therefore, we had to take the basic concept and make it more industry friendly.

Before we continue on with the discussion on how to develop an FMEA, let's first talk about why you would want to perform one in the first place. There really are two basic reasons to perform an FMEA in industry. The first and foremost is to make a legitimate business case to analyze one event vs. another. In other words, it creates the financial or business case to show a listing of all the events within a given organization or system and delineate in dollars and cents why you are choosing one issue over another. It allows the technical representative to speak in the language of business.

The second compelling reason is to focus the organization on what the most significant events are, so that quantum leaps in productivity can be made with fewer of the organization's resources being used. Experience again has shown that the Pareto Principle* works with such events just like it does in other areas. It goes

---

* RCFA Methods Course, Reliability Center, Inc., Hopewell, VA, 1985–1998.

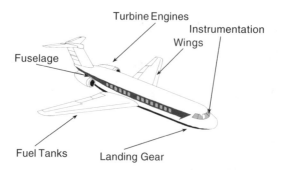

FIGURE 4.1 Aircraft subsystem diagram.

something like this: 20% or less of the undesirable events that we uncover by conducting an in-depth FMEA will represent approximately 80% of the losses for that organization. We may have heard this also called the 80/20 rule. We will talk more about the 80/20 rule later in this chapter.

As we mentioned before, the FMEA technique was developed in the aerospace industry and we will refer to this as the traditional FMEA method. Modifications are necessary to make the traditional FMEA more applicable to other organizations. Therefore, based on the modifications that we will explain in this chapter, we will call this technique the *modified* FMEA method. The key difference between the two methods is that the traditional method is probabilistic, meaning that it is looking at what could happen.

In contrast, the modified method looks only at historical events. We only list items that have actually happened in the past. For the modified method, we are not exactly as interested in what *might* happen tomorrow as we are in what *did* happen yesterday.

TRADITIONAL FMEA

Let's take a look at a simple example of both a traditional FMEA and a modified FMEA. Our intention is not to develop experts in traditional FMEA but to give a general understanding of how FMEA was derived. In the aerospace industry, we would perform an FMEA on a new aircraft that is being developed. So the first thing we might do is take that aircraft and break it down into smaller subsystems. So a typical aircraft would have many subsystems such as the wing assembly, instrumentation system, fuselage, engines, etc., as shown in Figure 4.1.

From there the analysis would look at each of the subsystems and determine what failure events might occur, and if they did, what their effects would be. Let's take a look at a simple example in Table 4.1.

In Table 4.1, we begin by looking at the turbine engine subsystem. We begin listing all of the potential failure modes that might occur on the turbine engines. In this case, we might determine that a turbine blade could fracture. We then ask what the effects on other items within the turbine engine subsystem might be. If the blade

TABLE 4.1
Traditional FMEA Sample

Subsystem	Mode	Effects on Other Items	Effects on Entire System	Severity	Probability	Criticality
Turbine engine	Cracked blade	If blade releases, it could fracture other blades	Loss of one engine, reduced power & control	8	0.02	0.16

was to release, it could fracture the other turbine blades. The effects on the entire system, or the aircraft as a whole, would be loss of the engine and reduced power and control of the aircraft. We then begin examining the severity of the failure mode. We will use a simple scale of 1 to 10 where 1 is the least severe and 10 is the most severe. We have simplified this for explanation purposes, but a traditional FMEA analyst would have specific criteria for what constitutes a severity of 1 through 10. In this example, we will say that losing a turbine blade would constitute a severity of 8. Now comes the probability rating. We would have to collect enough data to determine the relative probability of this occurrence based on the design of our aircraft. We will assume that the probability in this case is .02, or 2%. The last step is to multiply the severity times the probability to get a criticality rating. In this case, the rating would be calculated as shown in the following sample criticality equation:

$$\text{Severity} \times \text{Probability} = \text{Criticality}$$

$$8 \times .02 = .016$$

This means that this line item in the FMEA has a criticality rating of .016. We would then repeat this process for all of the failure modes in the turbine engines and all of the other major subsystems.

Once all of the items have been identified it is time to prioritize. We would sort our criticality column in descending order so that the largest criticality ratings would bubble up to the top and the smaller ones would fall to the bottom. At some point the analyst would make a cut specifying that all criticalities below a certain number are delineated as an acceptable risk, and all above need to be evaluated to determine a way to reduce the severity and, more important, the probability of occurrence.

Bear in mind that this is a long-term process. There is a great deal of attention given to determining all of the possible failure modes and even greater attention paid to substantiating the severity and probability. Thousands of hours are spent running components to failure to determine probability and severity. Computers, however, have helped in this endeavor, in that we can simulate many occurrences by building a computer model and then playing what-if scenarios to see what the effects would be.

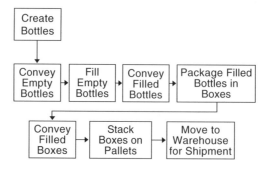

FIGURE 4.2 Sample lubricant plant.

TABLE 4.2
Modified FMEA Line Item Sample

Subsystem	Event	Mode	Frequency/Yr.	Impact	Total Loss
Fill empty bottles	Bottle stoppage	Bottle jam	1,000	$150	$150,000

MODIFIED FMEA

We do not have the time or resources in business and industry to perform a thorough traditional FMEA on every system. It does not even make economic sense to do so on every system. What we have to do is modify the FMEA process to help us to uncover the problems and failures that are currently occurring. This allows us to see what the real cost of these problems is and how they are really affecting our operation. Let's look at a simple example.

Consider that we are running a lubricants plant. In this plant we are doing the following:

1. Creating the plastic bottles for the lubricant
2. Conveying the bottles to the filling machine to be filled with lubricant
3. Conveying the filled bottles to the packaging process to be boxed in cases
4. Conveying the filled boxes to be put onto pallets
5. Moving the pallets to the warehouse where they await shipping

These steps are illustrated in Figure 4.2.

The next step is to determine all of the undesirable events that are occurring in each of our subsystems. For instance, if we were looking at the "fill empty bottles" subsystem, we would uncover all of the undesirable events related to this subsystem. Let's look at this simple example, shown in Table 4.2.

The idea is to delineate the events that have occurred that caused an upset in the fill empty bottles subsystem. In this case, one of the events would be a bottle stoppage. The mode of this particular event is that a bottle became jammed in the filling cycle. It occurs approximately 1000 times a year or about three times a day. The approximate impact for each occurrence is $150 in lost production. If we multiply the frequency times the impact for each occurrence, we would come to a total loss of $150,000 per year.

If we were to continue the analysis, we would pursue each of the subsystems delineating all of the events and modes that have caused an upset in their respective subsystems. The end result would be a listing of all the items that contribute to lost production and their respective losses. Based on that listing, we would select the events that were the greatest contributors to lost production and perform a disciplined root cause analysis (RCA) to determine the root causes for their existence.

Now that we understand the overall concept of FMEA, lets take a detailed look at the steps involved in conducting a modified FMEA. There are seven basis steps involved in conducting a modified FMEA:

1. Perform preparatory work
2. Collect the data
3. Summarize & encode results
4. Calculate loss
5. Determine the "significant few"
6. Validate results
7. Issue a report

STEP 1: PERFORM PREPARATORY WORK

As with any analysis, there is a certain amount of preparation work that has to take place. FMEA is no different, in that it also requires several up-front tasks. In order to adequately prepare to perform a modified FMEA, you must accomplish the following tasks:

- Define the system to analyze
- Define undesirable event
- Draw block diagram (use contact principle)
- Describe the function of each block
- Calculate the "gap"
- Develop preliminary interview sheets and schedule

Define the System to Analyze

Before we can begin generating a list of problems, we have to decide which system to analyze. This may sound like a simple task, but it does require a fair amount of thought on the analyst's part. When we teach this method to our students, their usual response is to take an entire facility and make it the system. This is a prescription for disaster. Trying to delineate all of the failures and/or problems in a huge oil refinery for instance would be a daunting task. What we need to do is localize the system down to one system within a larger system. For instance, a large oil refinery is composed of many operating units. There is a crude unit, fluid catalytic cracking unit (FCCU), coker, and many others. The prudent thing to do would be to select one unit at a time and make that unit the focus of the FMEA. For example, the crude unit would be the system to study and then we would break the crude unit into many subsystems. In other words, we should not bite off more than we can chew when selecting a system to study. We have seen many cases where the FMEA analysts first do a rough cut to see which area of the facility either constitutes a bottleneck or is incurring the greatest amount of expense.

Define Undesirable Event

This may sound a little silly, but we have to define exactly what an undesirable event is in our facility. During every seminar that we teach on this subject, we ask the students in class to write down their definition of an undesirable event at their facility. Just about every time, every student has a different definition. The fact is, if we are going to collect event data, everyone involved must be using a consistent definition. If we are collecting event data and there is no standardized definition, then everyone will give us their perceptions of what undesirable events are occurring in their work areas. For instance, if we ask a machine operator what undesirable events he sees, he will probably give us processing type events, a maintenance mechanic will probably give us machine-related events, whereas a safety engineer would probably give all of the safety issues. The dilemma here is that we lose focus when we do not have a common definition of what an undesirable event is.

The key to effectively defining an undesirable event is to make sure that the definition coincides with a particular business need. For example, if we are in a sold out position, then our definition should be based primarily around continuous production or limiting downtime. If we are not in a sold out position, then we might have to focus more on how to lower our costs in order to increase product margins. Let's take a look at some common definitions that we have run across over the years. Some are pretty good and some others are unacceptable. An undesirable event is:

- Any loss that interrupts the continuity of maximum quality production
- A loss of asset availability
- The unavailability of equipment
- A deviation from the status quo
- Not meeting target expectations
- Any secondary defect

The first definition — an undesirable event is any loss that interrupts the continuity of maximum quality production — is pretty good and one that we see and use quite frequently. Let's analyze this definition. In most manufacturing facilities, we often take our processes offline to do routine maintenance. The question is whether these planned shutdowns are undesirable events based on the first definition above. The answer is an emphatic *yes*! The definition states any loss that interrupts the continuity of maximum quality production is deemed an undesirable event. Even if we plan to take the machines out of service, it still interrupts the continuity of maximum quality production.

Now we are not saying that we should not take periodic shutdowns for maintenance reasons. All we are suggesting is that we look at them as undesirable events
so that we can analyze if there is any way to stretch out the intervals between each
planned shutdown (mean time between failure) and reduce the amount of time a
planned shutdown actually takes (mean time to restore). For instance, in many
industries we still have what we call *annual shutdowns*. How often do we have an
annual shutdown? Every year of course! It says so right in the name. Obviously, the
government and other legislative bodies regulate some shutdowns such as pressure
vessel inspections. But in many cases, we are doing these yearly shutdowns just
because the calendar dictates it. Instead of performing these planned shutdowns on
a time basis, maybe we should consider using a condition basis. In other words, let
the condition of the equipment dictate when a shutdown has to take place.

This idea of looking at planned shutdowns as an undesirable event is not always
obvious or popular. But if we are in a sold out position, we must look at anything
that takes us away from our ability to run 8760 hours a year at 100% throughput
rate. Now let's consider a different scenario. In many facilities we have spare
equipment, just in case the primary piece of equipment fails. It is sort of an insurance
policy for unreliability. In this scenario, if the primary equipment failed and the
spare equipment kicked in, would there be a loss that would interrupt the continuity
of maximum quality production? Providing the spare functions properly, the answer
here would have to be *no*. Because we had the spare equipment in place and
operating, we did not lose the production. This would mean that event would not
end up on our list because it did not meet our definition of an undesirable event.
This is also a hard pill for some of us to swallow. But that is the tough part about
focusing. Once we define what an undesirable event is, we must list only the events
that meet that definition.

Let's consider the definition, " an undesirable event is a deviation from the status
quo." This definition has many problems. The primary problem is whether a positive
deviation should be considered a failure. Probably not. How about the words *status
quo*? For one thing, status quo is far too vague. If we were to ask 100 people to
describe the status quo of the U.S. today, they would all give us a different answer.
Plus the fact that status quo does not always mean that things are good, it just says
that things are the way they are. That definition would make more sense rewritten
like this:

An undesirable event is a negative deviation from 1 million units per day.

So why bother with a definition? It serves multiple purposes. First of all, we
cannot perform a modified FMEA without it. But in our opinion, that is the least
important reason. The biggest advantage of an agreed upon definition is that it fosters
precise communication between everyone in the facility. It gets people focused on
the most important issues. In short, it focuses people on what is really important.

When we devise a definition of an undesirable event, we need to make sure that
it is short and to the point. We certainly would not recommend a definition that is
several paragraphs long. A good definition can and should be about one sentence.
Our definition should only address one business need at a time. For example, a

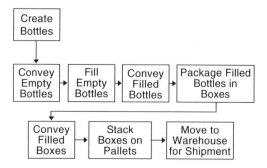

FIGURE 4.3 Block diagram example.

definition that states "An desirable event is anything that causes downtime, an injury, an environmental excursion, and/or a quality defect" is trying to capture too many items at one time, which in turn will cause the analysis to lose focus. If we feel the need to look at each of those issues, then we need to perform separate analyses for each of them. It may take a little longer, but we will maintain the integrity of the analysis focus.

Last but not least, it is important to get decision makers involved in the process. We would recommend having someone in authority sign off on the definition to give it some credence and clout. If we are lucky, the person in authority will even modify the definition. This will, in essence, create buy-in from that person.

Draw Block Diagram (Use the Contact Principle)

Now that we have defined the system to analyze and the definition of an undesirable event that is most appropriate, we have to create a simple flow diagram of the system being analyzed. This diagram will serve as a job aid later when we begin collecting data. The idea of a diagram is to show the flow of product from point A to point B. We want to list out all of the systems that come in contact with the product.

Let's refer back to our lubrication facility example. Each of these blocks in Figure 4.3 indicates a subsystem that comes in contact with the product. We use this diagram to help to graphically represent a process flow so that it is easy to refer to. Many facilities have such detailed drawings in order to comply with various government regulatory agencies. Oftentimes such diagrams are referred to as process flow diagrams (PFDs). If we have such diagrams already in our facilities, we are ahead of the game. If we do not, we must simply create a simple diagram like the one in Figure 4.3 to help represent the overall process. We will discuss how to use both the undesirable event definition and the contact flow diagram in the data collection phase.

Describe the Function of Each Block

In some cases, drawing the block diagram in itself is not enough of an explanation. We may possibly be working with some individuals that are not intimately aware

Potential = 1000 Donuts / Day

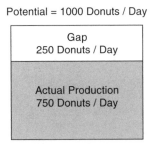

FIGURE 4.4 Sample gap analysis.

of the function of each of the systems. In these cases, it will be necessary for us to add some level of explanation for each of the blocks. This will allow those who are less knowledgeable in the process to participate with some degree of background in the process.

Calculate the "Gap"

In order to determine success, it will be necessary to demonstrate where we are as opposed to where we could be. In order to do this, we will need to create a simple gap analysis. The gap analysis will visually show where we currently are vs. where we could be. See Figure 4.4. For instance, let's assume that we have a donut machine that has the potential of making 1000 donuts per day, but we are only able to make 750 donuts per day. The gap is 250 donuts per day. We will use our modified FMEA to uncover all of the reasons that keep us from reaching our potential of 1000 donuts per day.

Develop Preliminary Interview Sheets and Schedule

The last step in the preparatory stage is to design an interview sheet that is adequate to collect the data consistent with your undesirable event definition and to set up a schedule of people to interview to get the required data. Let's look at the required data elements or fields. In every analysis we will have the following data elements:

- Subsystem — correlates to the blocks in our block diagram
- Event — the actual undesirable event that matches the definition we created earlier.
- Mode — the apparent reason that the undesirable event exists.
- Frequency per year — the number of times the mode actually occurs in a year's time.
- Impact per occurrence — the actual cost of the mode when it occurs. For instance, we will look at materials, labor, lost production, fines, scrap, etc. This data element can represent any item that has a determinable cost.
- Total loss per year — the total loss per year for each mode. It is calculated by simply multiplying the frequency per year by the impact per occurrence.

TABLE 4.3
Sample Modified FMEA Worksheet

Subsystem	Event	Mode	Frequency/Yr.	Impact Labor	Materials	Downtime	Total Loss
Area 1	Conveyor failures	Belts fail	104	$100	$25	$500	$65,000

In order to develop an effective interview sheet (see Table 4.3) we have to create it based on our definition. The first four columns (subsystem, event, mode, and frequency) are always the same. The impact column, however, can be expanded to include whatever cost elements that we feel are appropriate for the given situation. For instance, some do not include straight labor costs because we have to pay such a cost regardless. We will, however, include any overtime costs associated with the mode because we would not have incurred the expense without the event occurring.

The last item in the preparatory stage is to determine which individuals we should interview and to create a preliminary interview sheet to list all of the individuals to talk to in order to collect this event information. We will talk more about what types of people to interview in the next topic.

STEP 2: COLLECT THE DATA

There are a couple of schools of thought when it comes to how to collect the data that is necessary to perform a modified FMEA. On one side, there is the school that believes that all data can be retrieved from a computerized system within our organization. The other side believes that it would be virtually impossible to get the required data from an internal computer system because the data going into the system is suspect at best. Both sides are to some degree correct. Our data systems do not always give the precise information that we need, although they can be useful to verify trends that would be uncovered by interviewing people.

We will explore both of these alternatives in this chapter and the next. However, the analyst leading the modified FMEA will ultimately be responsible for making the decision as to whether the more accurate and timely data comes from the people or the existing information system. In this chapter we will continue on with the manual approach of collecting data from the raw source, the people. In Chapter 5, we will explore the data collection opportunities that are available from an enterprise reliability management system (ERMS), hence automating the effort.

It is recommended that when using the manual method of data collection (interviewing technique), we take a two-track approach. We begin collecting data from people through the use of interviews. We use the interviews very loosely as we will explain later. Once we have collected and summarized the interview data, we can use our existing data systems to verify financial numbers and see if the computer data supports the trends that we uncovered in our interviews. The numbers will not be the same, but the trends may very well be. So let's discuss how to go about collecting event data using an interview method.

Remember from our previous discussion, we developed two job aids. We had an undesirable event definition and a block diagram of the process flow. We are now going to use those two documents to help us structure an interview. We begin the interview by asking the interviewee to delineate any events that meet our definition of an undesirable event within a certain subsystem. This creates a focused interview session. As we said before, an interview generally has a kind of negative connotation. In order to gain employment we typically had to go through an interview, which is sometimes a stressful situation. We often watch TV police shows where a suspect is being interviewed (i.e., interrogated) in a dark smoky room. We would choose to make our interviews much more informal. Think of them more as an information-gathering session instead of a formal interview. This will certainly improve the flow of information.

Now who would be good candidates to talk to in an interview or discussion session? It is important to make sure that we have a good cross section of people to talk to. For instance, we would not want to talk to just maintenance personnel because we may only get maintenance-related items. So what we have to strive to do is interview across disciplines, meaning that we get information from maintenance, operations, technical, and even administrative personnel. Only then will we have the overall depth that we are looking for. There is also the question of what level of person we want to talk with. In most organizations there is a hierarchy of authority and responsibility. For instance, in a manufacturing plant there are the hourly or field-level employees who are primarily responsible for operating and maintaining the day-to-day operations to keep the products flowing. Then there is a middle supervisory level who typically supervise the craft and operator levels. Above the supervisory levels are the management levels who typically are looking at the operation from a more global perspective.

When trying to uncover undesirable events and modes, it makes sense to go to the source. This means talking to the people closest to the work. In most cases, this would describe the hourly workforce. They deal with undesirable events each and every day and are usually the ones responsible for fixing those problems. For this reason, we would recommend spending most of the interview time with this level. If we think about it, the hourly workforce is the most abundant resource, but rarely in our experience is it used to its fullest potential. Sometimes getting this wealth of knowledge is as easy as just asking for it. We are certainly not suggesting that we should not talk with supervisory level employees or above. They also have a vast amount of experience and knowledge of the operation. As far as upper level managers go, they usually have a more strategic focus on the operation. They may not have the specific information required to accomplish this type of analysis. There are exceptions to every rule, however. We once worked at a facility where the plant manager routinely would log into the distributive control system (DCS) from his home computer in the middle of the night to observe the actions of his operators. When they made an adjustment that he thought was suspect, he would literally call the operator in the control room to ask why they did what they did. Can we imagine trying to operate in such a micro-managed environment? Although we do not support this manager's practice, he probably would have much to offer in our analysis of process upsets because he had intricate knowledge of the process itself.

Another idea that we have found to be very useful when collecting event information is to talk to multiple people at the same time. This has several benefits. For one, when a single person is talking it is spurring something in someone else's mind. It also has a psychological effect. When we ask people about event information, it may be perceived as a witch hunt. In other words, they might feel like management is trying to blame people for the event. By having multiple interviewees in a session, it appears to be more of a brainstorming session than an interrogation.

The interviewing process, as we have learned over the years, is really an art form more than a science. When we first start to interview, we soon learn that it can sometimes be a difficult task. It is like golf, the more we practice proper technique the better the final results will be. An interview is nothing more than getting information from one individual to another as clearly and accurately as possible. To that end, here are some suggestions that will help us become more effective interviewers. Some of these are very specific to the modified FMEA process, but others are generic in that they can be applied to any interviewing session.

- Be very careful to ask the exact same lead questions to each of the interviewees. This will eliminate the possibility of having different answers depending on the interpretation of the question. Later we can expand on the questions, if further clarification is necessary. We can use our undesirable event definition and block flow diagram to keep the interviewees focused on the analysis.
- Make sure that the participants know what a modified FMEA is, as well as the purpose and structure of the interviews. If we are not careful, the process may begin to look more like an interrogation than an interview to the interviewees. An excellent way to make our interviewees comfortable with the process is to conduct the interviews in their work environments instead of ours. For instance, go to the break area or the shop area to talk to these people. People will be more forthcoming if they are where they are most comfortable.
- Allow the interviewees to see what we are writing. This will set them at ease since they can see that the information they are providing is being recorded accurately. Never use a tape recorder in a modified FMEA session because it tends to make people uncomfortable and less likely to share information. Remember, this is an information-gathering session and not an interrogation.
- If we do not understand what someone is telling us, let them use a pen to draw a simple diagram of the event for further understanding. If we still do not understand what they are trying to describe, then we should go out to the work area where the problem is so that we can actually visualize the problem.
- Never argue with an interviewee. Even if we do not agree with the person, it is best to accept what they are saying at face value and double check it with the information from other interviews. The minute we become argumentative, it reduces the amount of information that we can get from

that person. Chances are that person will not only not give us any more information, but will also alert others to the argument and they will not want to participate either.

- Always be aware of interviewees' names. There is nothing sweeter to people's ears than the sound of their own name. If you have trouble remembering, simply write the names down so that you can always refer to them. This gives any interview or discussion a more personal feel.

- It is important to develop a strategy to draw out quiet participants. There are many quiet people in our workforce who have a wealth of data to share but are not comfortable sharing it with others. We have to make sure that we draw out these quiet interviewees in a gentle and inquiring manner. We can use nominal group technique where we ask each of the people to whom we are talking to write their comments down on an index card and then compile the list on a flip chart. This gives everyone the same chance to have their comments heard.

- Be aware of body language in interviewees. There is an entire science behind body language. It is not important that we become expert in this area. However, it is important to know that a substantial portion of human communication is through body language. Let the body language talk to us. For instance, a person who sits back in a chair with arms firmly crossed may be apprehensive and not feel comfortable providing the information that we are asking for. This should be a clue to alter our questioning technique to make that person more comfortable with the situation.

- In any set of interviews, there will be a number of people who are able to contribute more to the process than some of the others. It is important to make a note of the extraordinary contributors so that they can assist us later in the analysis. They will be extremely helpful if we need additional event information for validating our finished modified FMEA, as well as assisting us when we begin our actual root cause analysis (RCA).

- Remember to use our undesirable event definition and block diagram to keep interviewees on track if they begin to wander off of the subject.

- We should try to never exceed one hour in interview sessions. This process can be very intensive and people can become tired and sometimes lose their focus. This is dangerous because it begins to upset the validity of the data. So as a rule, one hour of interview is plenty.

STEP 3: SUMMARIZE AND ENCODE DATA

At this stage, we have generated a vast amount of data from our interviews. We now have to begin summarizing this information for accuracy. Chances are that when we are conducting our interviews, we will be getting some redundant data. For instance, a person from the night shift might be giving us the same events that the day shift person gave us. So we have to be very careful to summarize the information and encode it properly so that we do not have redundant events and are essentially "double dipping."

TABLE 4.4
Logical and Illogical Coding Example

Subsystem	Failure Event	Failure Mode
	Logical Coding	
Area 6	Pump 102 failure	Bearing fails
Area 6	Pump 102 failure	Seal fails
Area 6	Pump 102 failure	Motor fails
	Illogocal Coding	
In Area 6	Pump 102 failure	Bearing break
Area (6)	Failure of CP-102	Seals
Area 6	Pump failure - 102	Failure of motor

The easiest way to collect and summarize the data is to input it into an electronic spreadsheet or database like Microsoft® Excel* and Microsoft® Access**. Of course you could do this manually with a pencil and paper, but if a computer is available, take the opportunity to use it. It will save many hours of frustration with performing the analysis manually. Once you have input all of the information into the spreadsheet, look for any redundancy. Something to always remember when inputting information into a computer is to use a logical coding system. Once you define what that logical coding system is, stick to it. Otherwise the computer will be unable to provide the results you are trying to achieve. Let's take a look at the example in Table 4.4 to help us understand what we mean by logical coding.

If we were to use the coding portrayed at the bottom of Table 4.4 we would get erratic results when we tried to sort based on any of the columns. Therefore, we have to strive to use a coding system like the one depicted in the top of the graphic, which will give the required result when using various sorting schemes.

Now, how can we eliminate the redundant information that is given in the interview sessions? The easiest way is to take the raw data from our interviews and input it into our spreadsheet program. From there we can use the powerful sorting capability of the program to help look for the redundant events. The first step is to sort the entire list by the subsystem column. Then within each subsystem, we will need to sort the failure event column. This will group all of the events from a particular area so that we can easily look for duplicate events. Once again, if we do not use logical coding this will not be effective. So we should strive to be disciplined in our data entry efforts.

Let's take a look at Table 4.5 for an example of how to summarize and encode events.

In this example, we are looking at the recovery subsystem and we have sorted by the recirculation pump failures. Four different people at four separate times gave

---

* Microsoft Excel is a registered trademark of the Microsoft Corporation, 1985–1998.
** Microsoft Access is a registered trademark of the Microsoft Corporation, 1989–1998.

TABLE 4.5
Example of Summarizing and Encoding Results

Subsystem	Event	Mode	Frequency	Impact
Recovery	Recirculation pump fails	Bearing locks up	12	12 hours
Recovery	Recirculation pump fails	Oil contamination	6	1 day
Recovery	Recirculation pump fails	Bearing fails	12	12 hours
Recovery	Recirculation pump fails	Shaft fracture	1	5 days

TABLE 4.6
Example of Merging Like Events

Subsystem	Event	Mode	Frequency	Impact
Recovery	Recirculation pump fails	Bearing problems	12	12 hours
Recovery	Recirculation pump fails	Shaft fracture	1	5 days

us these events. Do we see any redundancy? The easiest way to do this is to look at the modes. In this case we have two that mention the word *bearing*. The second is oil contamination. The interviewee was probably trying to help us out by trying to give us the cause of the event. But he also was talking specifically about the bearing. So in essence the first three events are really the same event. So we will have to summarize the three events into one. Table 4.6 shows a summary of the items.

STEP 4: CALCULATE LOSS

Calculating the individual modes is a relatively simple process. The idea here is to multiply the frequency per year by the impact per occurrence. So if we have a mode that costs $5,000 per occurrence and it happens once a month, then we have a $60,000 per year problem. We usually choose to use dollars as the measurement to accurately determine loss. We may find that using another metric is a more accurate measurement for our business. For instance, we may want to track pounds, tons, number of defects, etc. But if it is possible, we should try to convert our measurement into dollars. Dollars are the language of business and are usually the easiest to communicate to all levels of the organization.

Let's look at Table 4.7 for a few examples of calculating the loss.

In this example, we are simply multiplying the frequency per year by the impact per occurrence, which in this case is in number of units. In other words, when each of these modes occur, the impact is the number of units lost as a result. Notice that the last column is total loss in dollars. We simply multiply the number of lost units by the cost of each unit to give a total loss in dollars. That's all there is to it!

TABLE 4.7
Example of Calculating the Loss

Event	Mode	Frequency	Impact	Total Lost Units	Total Loss ($)
	Frequency × Impact = Total Loss				
Pump failure	Bearing problems	12	500	6,000	$180,000
Off spec. product	Wrong color	52	400	20,800	$624,000
Conveyor failures	Roller failures	500	50	25,000	$750,000

STEP 5: DETERMINE THE SIGNIFICANT FEW

We now have to determine which events out of all the ones we have listed are significant. We have all heard of the 80/20 rule, but what does it really mean? This rule is sometimes referred to as the Pareto Principle. The name Pareto comes from the early 20th century Italian economist who once said that, "In any set or collection of objects, ideas, people, and events, a *few* within the sets or collections are *more significant* than the remaining majority." This rule or principle demonstrates that in our world, some things are more important than others. Let's look at a few examples of this rule in action:

- Banking industry — In a bank, approximately 20% or less of the customers account for approximately 80% or more of the assets in that bank.
- Hospital industry — In a hospital, approximately 20% or less of the patients get 80% or more of the care in that hospital.
- Airline industry — Twenty percent or less of airline passengers account for 80% or more of the revenues for the airline.

The fact is that it works in industrial applications as well. Throughout our years of experience, and our clients' as well, the rule holds true. Twenty percent or less of the identified events typically represent 80% or more of the resulting losses. This is truly significant if you think about it. It says that if we focus on and eliminate the 20% of the events that represent 80% of the losses, we will achieve quantum leaps in productivity in a relatively short period of time. It just makes common sense!

Think about how the rule applies to everyday living. We probably all are guilty of wearing 20% or less of the clothes in our closet 80% or more of the time. We probably all have a tool box in which we use 20% or less of the tools 80% or more of the time. We spend all that money on all those exotic tools and then we never even use them. We are all guilty of this! The rule even applies to business. Take for instance a major airline as described above. It is not the once-a-year vacationer who generates most of their revenue. It is the guy who flies every Monday morning and returns every Friday afternoon. So it makes sense that very few of their customers represent most of their revenue and profits. That is why we see all of the new frequent flier programs. They know whom they have to cater to.

Let's take a look at the following steps and the corresponding example in Table 4.8 to determine exactly how to take a list of events and narrow it down to the "significant few."

TABLE 4.8
Sample Modified FMEA Worksheet

Subsystem	Event	Mode	Frequency	❶ Impact	Total Loss
Subsystem A	Event 1	Mode 11	30	$40,000	$1,200,000 ❺
Subsystem A	Event 2	Mode 7	4	230,000	920,000
Subsystem B	Event 3	Mode 1	365	1,350	492,750
Subsystem A	Event 2	Mode 5	10	20,000	200,000
Subsystem A	Event 2	Mode 8	10	10,000	100,000
Subsystem B	Event 5	Mode 6	35	2,500	87,500 ❹
Subsystem B	Event 4	Mode 4	1000	70	70,000
Subsystem A	Event 4	Mode 12	8	8,000	64,000
Subsystem B	Event 6	Mode 10	6	8,000	48,000
Subsystem C	Event 4	Mode 13	4	7,500	30,000
Subsystem B	Event 4	Mode 9	10	2,500	25,000
Subsystem A	Event 1	Mode 2	12	2,000	24,000
Subsystem A	Event 1	Mode 3	9	2,500	22,500
Subsystem C	Event 6	Mode 14	6	3,500	21,000
Total loss					**$3,304,750** ❷
Significant few losses (Total loss × .80)					**$2,643,800** ❸

1. Multiply the frequency column times the impact column to get a total annual loss figure.
2. Sum the total annual loss column to obtain a global total loss figure for all the events in the analysis.
3. Multiply the global total loss figure from Step 2 by 80%, or .80. The result will be the "significant few" losses amount.
4. Sort the total loss column in descending order so that the largest events bubble up to the top.
5. Sum the total loss amounts from biggest to smallest until you reach the "significant few" loss amount.

In order to get the maximum effect it is always wise to present this information in alternate forms. The use of graphs and charts will help to effectively communicate this information to others. Figure 4.5 shows a sample bar chart that takes the spreadsheet data above and converts it into a more understandable format.

STEP 6: VALIDATE RESULTS

Although our analysis is almost finished, there is still more to accomplish. We have to verify that our findings are accurate. Our modified FMEA total should be relatively close to the gap that we defined in the preparatory phase. The general rule is plus or minus 10% of the gap.

If we are way under that gap, we have either missed some events, undervalued them, or we do not have accurate gap (actual vs. potential). If we were to overshoot

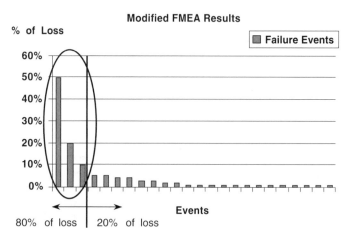

FIGURE 4.5 Sample bar chart of modified FMEA results.

the gap, we probably did not do a good job at removing the redundancies, or we have simply overvalued the loss contribution.

At a minimum we must double-check our "significant few" events to make sure we are relatively close. We do not look for perfection in this analysis simply because it would take too long to accomplish, but we do want to be in the ballpark. This would be a good opportunity to go to our data sources like our maintenance management system or our distributive control system (DCS) to verify trends and financial numbers. Incidentally, if there is ever a controversy over a financial number, it would be prudent to use numbers that the accounting department deems accurate. Also, it is better to be conservative with financials to not risk losing credibility because of an exaggerated number. The numbers will be high enough on their own without any exaggeration. Other verification methods might be more interviews or designed experiments in the field to validate interview findings. All in all, we want to be comfortable enough to present these numbers to anyone in the organization and feel that we have enough supporting information to back them up.

STEP 7: ISSUE A REPORT

Last but certainly not least, we have to communicate our findings to decision makers so that we can proceed to solve some of these pressing issues. Many of us falter here because we do not take the time to adequately prepare a thorough report and presentation. In order to gain maximum benefit from this analysis, we have to prepare a detailed report to present to any and all interested parties. The report format is based primarily on style, whether that be our personal style or a mandated company reporting style. We would suggest the following items to be included in the report in some form or fashion.

- Explain the analysis — Many of our readers may be unfamiliar with the modified FMEA process. Therefore, it is in our best interest to give them a brief overview of what a modified FMEA is and what its goal and benefits are. This way, they will have a clear understanding of what they are reading.

- Display results — Provide several charts to represent the data that our analysis uncovered. The classic bar chart demonstrated earlier is certainly a minimal requirement. In addition to supporting graphs, we should provide all the details. This includes any and all worksheets compiled in the analysis.

- Add something extra — We can be creative with this information to provide further insight into the facility's needs by determining other areas of improvement other than the "significant few." For instance, we could break out the results by subsystem and give a total loss figure for each subsystem. The manager of that area would probably find that information very interesting. We could also show how much the facility spent on particular modes like bearings or seals. This might be interesting information for the maintenance manager. We must use our imagination as to what we think is useful, but by using the querying capabilities of our spreadsheet or database, we can gleam any number of interesting insights from this data.

- Recommend which event to analyze — We could conceivably have a couple dozen events from which our "significant few" list is composed. We cannot work on all of them at once, so we must prioritize which events should be analyzed first. Common sense would dictate going after the most costly event first. On the surface this sounds like a good idea, but in reality we might be better off going after a less significant loss that has a lesser degree of complexity. We like to call these events the "low hanging fruit." In other words, go after the event that gives the greatest amount of payback with the least amount of effort.

- Give credit where credit is due — We must list each and every person who participated in the FMEA process. This includes interviewees, support personnel, and the like. If we want to gain their support for future analyses, then we have to gain their confidence by giving them credit for the work they performed. It is also critical to make sure that we feed the results of the analysis back to these people so they can see the final product. We have seen any number of analyses fail because participants were left out of the feedback loop.

That is all there is to performing a thorough modified FMEA. As we mentioned before, this technique is a powerful analysis tool, but it is also an invaluable sales tool in getting people interested in our projects. If we think about it, it appeals to all parties. The people who participated will benefit because it will help eliminate some of their unnecessary work. Management will like it because it clearly demonstrates what the return on investment will be if those events or problems are resolved. Keep in mind that we may not be able to implement this process as cleanly as we have tried to specify in this book. All that is important is that we take the overall concept and apply it to our given situation so that it works best for us.

5 Enterprise Reliability Asset Management Systems (ERMS): Automating the Modified FMEA Process

Up until now, we have discussed the manual interview method of collecting event data to determine the best candidates for RCA. Now, let's consider automating the process of event data collection. When we talk about automating data collection, we are really discussing how to collect event data on a day-to-day basis using modern data collection and analysis tools. When we employ sophisticated data analysis techniques, we actually have the ability to view the data in a way that turns raw data into actionable information.

In this chapter we will discuss what is needed to implement a comprehensive event-recording data system. Below are the core activities that need to be established to enable the automated data analysis infrastructure:

- Establish key performance indicators (KPIs)
- Determine your event data needs
- Establish a workflow to collect the data
- Employ a comprehensive data collection system
- Analyze the digital data

ESTABLISH KEY PERFORMANCE INDICATORS (KPIs)

Tom Peters once said, "You can't improve what you cannot measure." If you think about it for a minute, it makes a lot of sense. We have been exposed to KPIs since we were very young. From the moment we are born, we are weighed and measured and then we are compared to standards to see what percentile we are in. As we grow and get into school, we are exposed to another set of KPIs, the infamous report card. The report card allows us to compare our performance against our peers' or to some other standard. An example that many people can certainly relate to is the use of a scale to measure the progress of a diet. We probably would not be very successful if we did not know where we started and what progress we were making week by week.

We all need a scoreboard to help us determine where we started and where we are at any given time. This certainly applies to measuring the performance of a maintenance and reliability organization. We need to know how many events occur in a given month, on a specific class of equipment, etc. Not until we know what KPIs will effectively measure our maintenance and reliability efforts can we begin to establish which opportunities will afford the greatest returns.

With all of that said, we would like to provide a word of caution. Be very careful to diversify your KPI selections. While a report card in school is a good measurement of a student's performance, it still does not provide a complete picture of the individual student. It is only one data point! Some students perform better on written tests while other students excel in other ways. We need to be careful to make sure that we employ a set of KPIs that most accurately represents our performance. That means we need many different metrics that look at different areas of performance so we can get a complete picture.

So let's take a look at a few common KPIs that can be employed to give us an understanding of our maintenance and reliability performance.

- MTBF (mean time between failure) — This is a common metric that has been used for many years to establish the average time between failures. Although it can be calculated in different ways, it is primarily looking at the total runtime of an asset and dividing by the total number of failures for that asset.

$$\text{Total Runtime/Number of Events} = \text{MTBF}$$

  This is a good metric because it is easy for people to understand and relate to and is common throughout industry. If a good data collection system is employed, the calculation is fairly easy to perform. The drawback is it does not give an indication of actual business impact on the facility. However, we can make the conclusion that if MTBF increases, the facility is performing better and providing a positive business impact.
- Number of events — This metric will simply measure the volume of events that occurs on a variety of dimensions. Those dimensions are typically process units, equipment classes (e.g., pumps), equipment types (e.g., centrifugal), manufacturer, and a host of others. This metric is closely related to MTBF as it is the denominator for the calculation. It can also be an accurate reflection of a facilities maintenance and reliability performance. It does however have the same pitfalls as the MTBF calculation.
- Maintenance cost — This metric simply measures the number of maintenance dollars that are expended on rectifying the consequence of an event. This is typically the sum of labor and material cost (including contractor costs). This metric is also employed across many different dimensions like equipment, areas, manufacturers, etc. This metric is a better business metric as it shows some of the financial consequences of the event. It also has some drawbacks, as it does not totally reflect the complete financial consequence of the event. It does not cover the lost opportunity (e.g., downtime) associated with the event. As we all know, the cost of downtime is much greater than the cost of maintenance on a dramatic downtime event.

- Availability — This metric is useful to determine how available a given asset or set of assets has been historically. In a 24/7 operation, the availability calculation (shown in equation 5.1) is simply the entire year's potential operating time minus downtime divided by total potential operating time.

$$\frac{8760 \ (\text{Total hours in a year}) - 32 \ (4 \text{ events of 8 hours each})}{8760 \ (\text{total hours in a year})} \tag{5.1}$$

Availability $= 99.63\%$

This calculation can be modified in many ways to fit a specific business need. Although this metric is a good reflection of how available the assets were in a given time period, it provides absolutely no data on the reliability or business impact of the assets.

- Reliability — This metric can be a better reflection of how reliable a given asset is based on its past performance. In the availability example above, we had an asset that failed four times in a year, resulting in 32 hours of downtime. The availability calculation determined that the asset was available 99.63% of the time. This might give the impression of a highly reliable asset. But if we use the reliability calculation shown in equation 5.2 we would get a much different picture.

Reliability $= e^{-\lambda t}$

Natural logarithmic base: $e = 2.718$

Failure rate: $\lambda = \dfrac{1}{\text{MTBF}} = \dfrac{1}{91}$

Mission time: $t = 365 \ (\text{days})$ $\tag{5.2}$

Reliability $= e^{-\lambda t}$ $\boxed{\begin{array}{l} e = 2.718 \\[2mm] \lambda = \dfrac{1}{\text{MTBF}} = \dfrac{1}{91} \\[2mm] t = 365 \ (\text{days}) \end{array}}$

$= 2.718^{-\lambda t}$

$= 2.718^{-\frac{1}{91}(365)}$

$= 2.718^{-4.0109}$

$= 1.81\%$

The fact of the matter is, an asset that fails four times per year is extremely unreliable and the likelihood of that asset reaching a mission time of one year is highly unlikely even though its availability is very good.

These are only a few common KPIs. As you can imagine, there is an array of metrics that can be used to help measure the effectiveness of a maintenance and reliability organization.

DETERMINING OUR EVENT DATA NEEDS

Once we have satisfactorily determined our performance metrics, it is time to determine the data required to accurately report on those metrics. Our data requirements will vary depending on our selection of KPIs, so we will provide some common data requirements to satisfy the more common metrics like the ones listed above.

Since we are focused on collecting event data, it is important to repeat what we discussed in the manual method. The definition of event is still critical whether we are performing modified FMEA manually or with an automated collection system. This definition is critical to the process and is typically the place where efforts like these become unsuccessful. As we might imagine, it is very difficult to collect data on something like events when the term has not been fully defined. What might be an event to you might not be considered an event to someone else! So follow some good advice and accurately define the event for your organization and then communicate that definition to all the relevant data collectors.

So what kind of data should be collected when an event occurs? Table 5.1 shows common data items that should be collected for any event.

The listing is by no means complete, but it is a good basis for getting a good event reporting system off the ground. Most of the KPIs we listed earlier in this chapter could be calculated with data in this list.

TABLE 5.1
Common Data Items to Collect for Any Event

Data Item	Description	Importance
Functional location	The functional location is typically a "smart" ID that represents what function takes place at a given location. (Pump 01-G-0001 must move liquid X from point A to point B.)	High
Asset ID	The asset ID is usually a randomly generated ID that reflects the asset that serves the functional location. The reason for a separate asset ID and functional location is that assets can move from place to place and functional locations never move. This is the reason we need to identify both event records to distinguish whether the problem is associated with the location or the asset itself.	High
Event date	This is the date that the event was first observed and documented.	High
Equipment category	This is the "high level" equipment that failed (e.g., rotating equipment).	High
Equipment class	This is the actual class of equipment that failed (e.g., pump).	High

TABLE 5.1 (continued)
Common Data Items to Collect for Any Event

Data Item	Description	Importance
Equipment type	This is the actual type of equipment that failed (e.g. centrifugal).	Medium
Unit or area	This uniquely identifies where the event took place within the facility (e.g., Unit 01 — crude unit).	High
Failed component	This is the actual component that was identified as causing the asset to lose its ability to serve (e.g., bearing).	High
Event mode	This is the mode or manner in which the component failed. This is sometimes subjective and in some cases difficult to determine without proper training and analysis (e.g., fatigue or erosion).	
Model number	This is the manufacturer model number of the asset that failed.	Medium
Material cost	This is the total maintenance expenditure on materials to rectify the event. This could be company or contractor cost.	High
Labor cost	This is the total maintenance expenditure on labor to rectify the event. This could be company or contractor cost.	High
Total cost	This is the total maintenance expenditure to rectify the event. This could be company or contractor cost.	High
Lost opportunity cost	This is the business loss associated with not having the assets in service. There is only a loss when an asset fails to perform its intended function and there is no spare asset or capability to make up the loss.	High
Other related costs	These are costs that might be incurred that do not relate directly to maintenance or lost opportunity (e.g., scrap, disposal, rework, fines, etc.).	High
Out of service date/time	This is the date/time that the equipment was actually taken out of service.	High
Maintenance start date/time	This is the date/time that the equipment was actually being worked on by maintenance.	Medium
Maintenance end date/time	This is the date/time that the equipment was actually finished being worked on by maintenance.	Medium
In service date/time	This is the date/time that the equipment was actually put back into service.	Medium

ESTABLISH A WORKFLOW TO COLLECT THE DATA

We do not want to minimize the difficultly related to collecting event data on a regular basis. The fact is that collecting accurate event data is extremely difficult to do. Event data is different from other types of data. Take process data, for instance, which is automatically being captured in a disciplined and consistent manner through the use of a distributive control system (DCS). The data is automatically captured with very little human interaction.

Event data, on the other hand, is very dependent on a variety of people collecting data in a uniform way. For instance, what one person might view as a pump event might actually be a motor (driver) issue. So how do we ensure that the data is compiled in a uniform manner?

First, we need to educate all stakeholders in the need for accurate data collection. In today's busy work environment we are constantly asked to collect an array of data. The problem with this approach is most people have no idea how the data they are being asked to collect is actually used. When this happens we begin to see entries in the computerized maintenance management system (CMMS) stating: "Pump broke, fixed it." This obviously gives no detail into the events and provides no opportunity to summarize the data for useful decision making. So before we ask anyone to collect data, we need to educate them in how the data will be used to make decisions.

The second step in the process is related to the first in that we need to develop definitions and codes to support the event data collection effort. This means that we need to determine common event codes for our equipment events and then educate our data collectors in the definition of these codes. A great way to do this is through the use of scenarios. The group of data collectors is presented with the various codes and their definitions. They are then subjected to a variety of event scenarios to test how they would use the codes in a variety of common situations.

Last, but not least, a comprehensive workflow will need to be established to collect the data described above. Essentially, an array of "W" questions need to be formulated and answered. For instance:

- Who will collect the data?
- What data is important?
- When will the data be collected?
- Where will it be stored?
- Who will verify the data?
- Who will enter the data?

We will answer many of these workflow questions when we discuss data collection systems. As a prelude to this, what many people do is try to use their CMMS system as the initial workflow to collect some of the data, and then devise a supplemental workflow to get the remaining data items. This is certainly one method and may be one of the most effective because some key reliability data is being generated through the use of the maintenance system.

EMPLOY A COMPREHENSIVE DATA COLLECTION SYSTEM

To truly automate the modified FMEA process, we need to use powerful data collection and analytical tools. Database technology has reached the point where different types of data systems can easily "talk" to each other so that a wide variety of data can be collected, summarized, and analyzed to allow analysts to make informed decisions.

TABLE 5.2
Common Data Fields

Asset ID	Maintenance start date/time
Functional location	Labor cost (in-house/contractor)
Manufacturer	Material cost (in-house/contractor)
Model number	Total work order cost
Event date	Unit
Failed component(s)	Equipment type

We are going to discuss a method for transferring data from existing computer-ized maintenance management systems (CMMS) into an enterprise reliability man-agement system or ERMS.* Before we discuss the interface between CMMS and ERMS, let's discuss the role of both of these system in the operation of the facility.

A CMMS is designed to assist maintenance personnel in the execution of work. The main function of this system is to automate the process of getting maintenance tasks completed in the field. This includes things like generating work requests, prioritizing work, planning and scheduling, managing materials, and finally, actually executing the work. CMMSs by nature are transaction-based systems, since many transactions have to take place to completely execute a maintenance event. There are many efficiencies that are gained by automating the maintenance workflow. Hence, most asset-intensive companies have implemented such systems to achieve the many benefits.

Although a CMMS provides a variety of benefits, it was not designed to be an analytical system to provide decision support to reliability and maintenance analysts. It does, however, offer a variety of good data that can be used to perform reliability analysis. For instance, every work order will delineate the asset ID and location of the maintenance event, the date the asset came out of service, and the components that we used to repair the asset. There is obviously much more than this, but those items alone can be extremely valuable in determining event probabilities and even optimizing preventive maintenance activities.

An enterprise reliability management system (ERMS) is not designed to handle maintenance workflow and transactions but to take that data and a variety of other data to create actionable information in which to improve the overall reliability and availability of the facility. These tools might contain extensive data manipulation tools, statistical analysis tools like Weibull Analysis, root cause analysis (RCA), risk based inspection (RBI), and many others. We will focus our discussion on how an ERMS can be a valuable aid to helping root cause analysts determine the best opportunities for analysis.

So what data can we use from a CMMS that would help an ERMS determine where the best opportunities for analysis might be? Table 5.2 contains some of the common data fields that would be useful in this type of analysis.

---

* ERMS is a registered trademark of Meridium, Inc., Roanoke, VA. (www.meridium.com).

This data is a solid starting point for performing modified FMEA for root cause events. The next step is to transfer this data into an ERMS so that the data can be supplemented with additional data about the event and then be "sliced and diced" to determine the opportunities.

In order to make use of this important data, the data must be somewhat easy to find and manipulate. Having worked with reliability and maintenance analysts for many years, I have seen a number of "homegrown" reliability management systems. I am sure that you too can attest to such systems. For example, what happens when a reliability engineer cannot seem to acquire the data needed to do the job? They build it themselves! They miraculously go from capable engineer to software developer. I am sure you have seen some of these masterpieces. They build them in spreadsheets, desktop databases, or even using full-blown development tools. Although these homegrown systems serve a valuable purpose for their creators, they have many pitfalls for an organization. For one, the data may or may not be accurate. Since the data is typically collected by a handful of users, it may not truly reflect the overall reality. The data may not be properly event coded so it becomes extremely difficult to analyze. The main problem with these homegrown solutions is that the data is not accessible to all the stakeholders who need it.

An ERMS is designed to interface with existing data sources like CMMS, PdM system, process systems, and a variety of others. See Figure 5.1. This ensures that the data is accurate and is kept up to date, as the interface keeps the system continually in sync. This is critical because it allows the data to be collected once and used for a variety of purposes. An ERMS is a secured system so you know that the data is protected. The most important purpose of an ERMS is to provide the value-added analysis tools to turn existing maintenance and reliability data into actionable information.

Let's move on to the area of analyzing your digital data.

ANALYZE THE DIGITAL DATA

The tool of choice to perform modified FMEA is the Pareto chart. Just to recap, a Pareto chart is simply a way to delineate the significant items within a collection. In our case, it will help us determine the few significant issues that represent the majority of the losses within a facility. The Pareto chart can be used on a variety of metrics depending on the need. For instance, some users might simply use maintenance cost as the only measure to determine whether an RCA needs to be initiated. Others might want to compile all the costs associated with an event, namely lost opportunity costs for not producing. Still others might be more interested in mean time between event (MTBF). The assets with the lowest MTBF might be the best candidates for RCA. The advantage of using an automated approach to modified FMEA is that the analyst can look for opportunities using a variety of metrics and techniques.

Today there are some powerful technologies to view and analyze data. One of the best for performing modified FMEA via Pareto charts is a technology called online analytical processing (OLAP). This technology allows users to view data with a variety of dimensions and measures. For instance, suppose you wanted to know

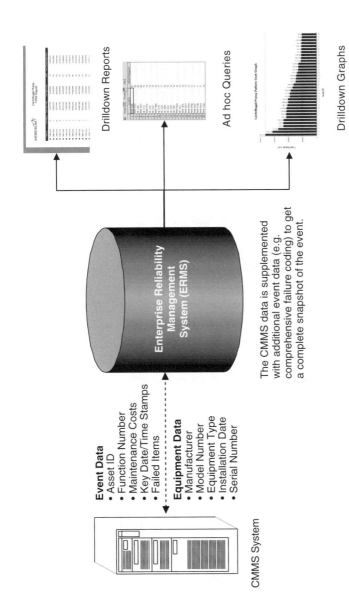

FIGURE 5.1 CMMS/ERMS interface.

which unit within your plant was responsible for the greatest maintenance expenditures. Once you knew that, the next obvious question might be which pieces of equipment were most responsible for that. To go even deeper, you might want to know what the component was that caused most of that expense. With OLAP tools, you can use powerful drill-down capability to do this type of analysis. Figures 5.2, 5.3, and 5.4 are a series of charts demonstrating these dynamic Pareto charts.

The use of OLAP makes sophisticated data mining easy for end users. It allows users to see what they want to see in the form that is the most useful for them.

Although OLAP is an incredible tool for dynamic modified FMEA, other tools might be useful as well. Some users might like to see the data presented in a particular format. For instance, suppose there is a corporate reporting standard that needs to be adhered to. If this was the case, the use of preformatted reports might make the most sense. Reports are useful for presenting predetermined metrics that are updated every time the particular report is run. Figure 5.5 is an example of a pump event count and maintenance cost report.

To allow for complete flexibility for data analysis, an ERMS would provide a comprehensive tool to perform ad hoc query ability. A query is simply a way to extract the information we need from the database. This is commonly done using the structured query language (SQL). SQL is the syntax, or language, needed to get the relevant data from the database. SQL is not something most analysts are intimately familiar with. So the ERMS must provide a highly flexible query tool that does not require the end user to know anything about SQL. Figures 5.6 and 5.7 provide an example of a query designed to determine the MTBF for a collection of pumps.

This only shows the surface of what can be accomplished when we automate modified FMEA. There are far more sophisticated statistical methods that can be employed. Our advice, however, is to start with the basics and slowly move into more sophisticated methods.

By automating modified FMEA, the users have a dynamic tool that allows them to look at opportunities in a variety of different ways. As business conditions change, then so can the opportunities. The key is to consistently collect the right data on a day-to-day basis.

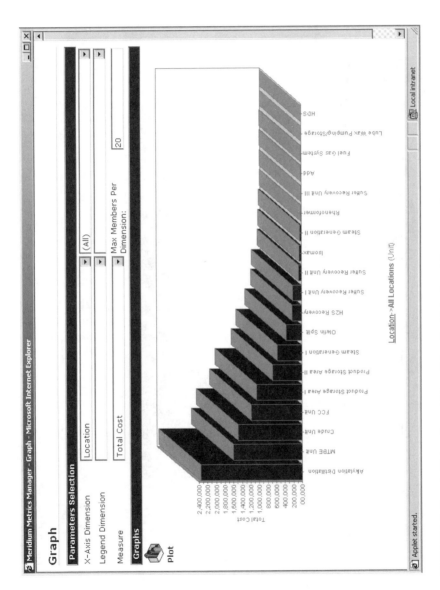

FIGURE 5.2 Step 1: Determine which unit has the highest maintenance cost.

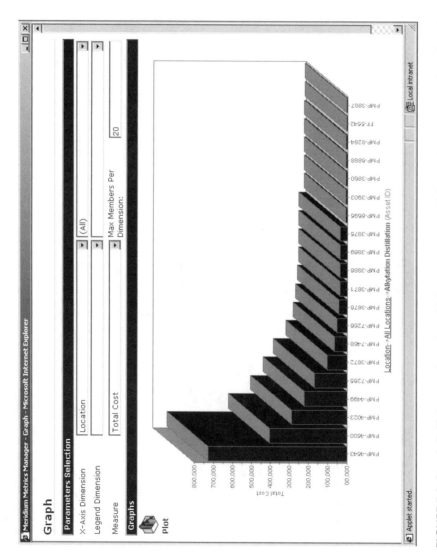

FIGURE 5.3 Step 2: Drill down to determine which assets represent highest maintenance cost from the unit with the highest cost (i.e., alkylation distillation).

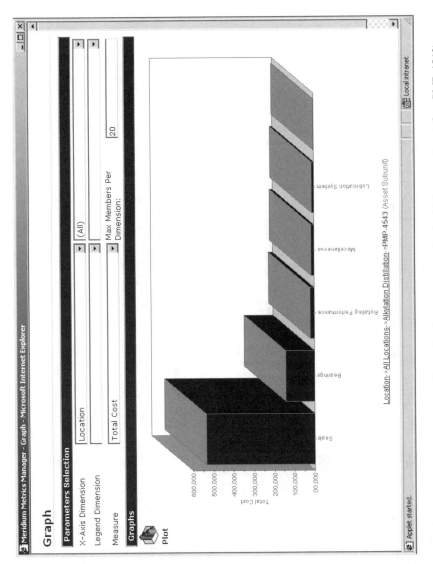

FIGURE 5.4 Step 3: Drill down to determine the components for the highest asset cost (i.e., PMP-4543).

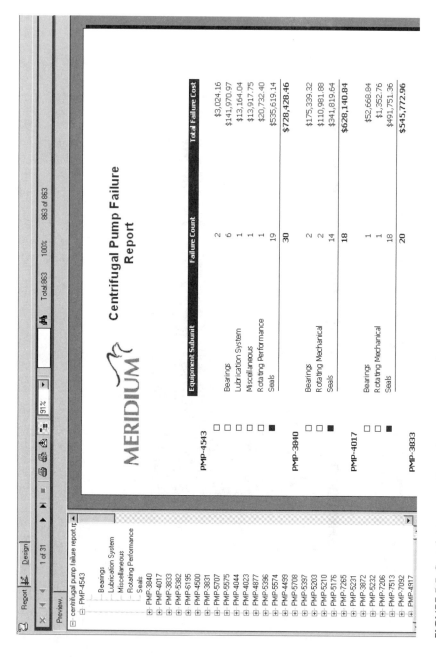

FIGURE 5.5 Sample pump event report.

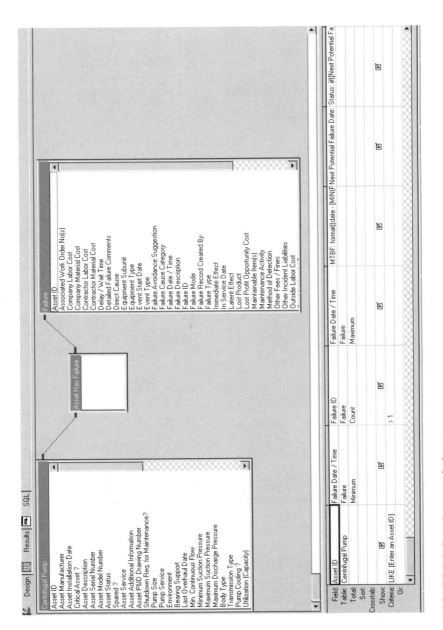

FIGURE 5.6 Sample query designer.

Asset ID	MIN(Failure Date / Time)	COUNT(Failure ID)	MAX(Failure Date / Time)	MTBF	Next Potential Failure Date	Status
FF-5542	1999-02-08	3	2000-06-07	319	2001-04-22	Overdue
PMP-1001	1998-10-15	4	2000-11-29	269	2001-08-25	Overdue
PMP-3121	2000-09-12	2	2000-10-18	188	2001-04-24	Overdue
PMP-3301	1998-12-17	3	2000-07-27	337	2001-06-29	Overdue
PMP-3582	1998-11-08	2	2000-04-02	525	2001-09-09	Overdue
PMP-3819	2000-03-14	3	2000-12-14	186	2001-06-18	Overdue
PMP-3820	1998-11-03	5	2000-06-20	211	2001-01-17	Overdue
PMP-3821	1998-11-19	3	2000-07-13	346	2001-06-24	Overdue
PMP-3822	1998-11-29	3	2000-07-13	343	2001-06-21	Overdue
PMP-3823	1999-06-06	5	2000-09-19	168	2001-03-06	Overdue
PMP-3825	1999-02-08	3	1999-09-28	319	2000-08-12	Overdue
PMP-3827	1998-11-30	2	2000-09-14	514	2002-02-10	
PMP-3830	1999-09-30	3	2001-01-17	241	2001-09-15	Overdue
PMP-3831	1991-09-16	15	1999-12-19	244	2000-08-19	Overdue
PMP-3833	1994-05-04	7	1997-10-29	386	1998-11-19	Overdue
PMP-3834	1999-03-08	3	2000-12-20	310	2001-10-26	
PMP-3839	1999-10-24	2	2000-07-10	350	2001-06-25	Overdue
PMP-3840	1991-01-10	18	2000-06-23	217	2001-01-26	Overdue
PMP-3848	1998-12-14	2	1999-11-01	507	2001-03-22	Overdue
PMP-3849	2000-01-25	2	2000-11-30	304	2001-09-30	
PMP-3851	1998-07-29	3	2000-11-07	384	2001-11-26	
PMP-3856	1999-11-01	2	2000-03-21	346	2001-03-02	Overdue
PMP-3859	2000-01-13	2	2000-03-07	310	2001-01-11	Overdue
PMP-3860	1999-05-10	3	2000-08-28	289	2001-06-13	Overdue
PMP-3868	1998-11-02	3	2000-11-14	352	2001-11-01	
PMP-3869	1998-09-20	4	1999-09-29	275	2000-06-30	Overdue
PMP-3871	1992-03-09	5	1999-11-30	697	2001-10-27	
PMP-3872	1999-03-07	2	2000-09-04	466	2001-12-14	
PMP-3874	1998-11-29	2	2000-02-15	515	2001-07-14	Overdue
PMP-3875	1996-06-22	3	2000-09-04	640	2002-06-06	
PMP-3876	1998-12-17	5	2000-12-04	202	2001-06-24	Overdue

FIGURE 5.7 Sample MTBF query results.

6 The PROACT® RCA Method

The term PROACT has recently come to light to mean the opposite of react. This may seem to be in conflict with the use of PROACT® as a root cause analysis tool. Normally when we think of root cause analysis, the phrase *after-the-fact* comes to mind. After, because by RCA's nature, an undesirable outcome must occur in order to spark action. So how can RCA be coined *proactive*?

In Chapter 4 on modified failure modes and effects analysis (FMEA), we clearly outlined a process by which to identify which failures or events were actually worth performing RCA on. We learned from this prioritization technique that, generally, the most important events to analyze are *not* the sporadic incidents, but rather the day-to-day chronic events that sap our profitability.

The RCA tools can be used in a reactive fashion and/or a proactive fashion. The RCA analyst will ultimately determine this. When we use RCA only to investigate incidents that are defined by regulatory agencies, then we are responding to the field. This is strictly reactive. However, if we use the FMEA tools described previously to prioritize our efforts, we will uncover events that many times are not even recorded in our computerized maintenance management systems (CMMS) or the like. This is because such events happen so often that they are no longer an anomaly, they are a part of the job. They have been absorbed into the daily routine, the norm. By uncovering such events and analyzing them, we are being proactive because unless we look at them, no one else will.

The greatest benefits from performing RCAs will come from the analysis of chronic events, hence using RCA in a proactive manner. We must understand that oftentimes we are getting sucked into the "paralysis by analysis" trap and end up expending too many resources to attack an issue that is relatively unimportant. We also at times refer to these as the "political failures of the day." Trying to do RCA on everything will destroy a company. It is overkill and companies do not have the time or resources to do it effectively. This is why we advocate doing true RCA as the "Significant Few" and using a less intensive approach (Failure Analysis/Problem Solving) for our day-to-day continuous improvement strategy. See Figure 6.1.

Understanding the difference between chronic and sporadic events heightens awareness of which data collection strategy will be appropriate for the event being analyzed. The key advantage with chronic events, if there can be one, is their frequency of occurrence, which often falls into a pattern. An analyst studying chronic events is like the detective stalking a serial killer, looking for a pattern to establish where the next crime might logically take place so he can prevent its occurrence. A chronic event will likely happen again within a certain time frame, and we may be able to plan for its occurrence and capture more data at that point in time. We will discuss this more when we go over verification techniques in Chapter 8.

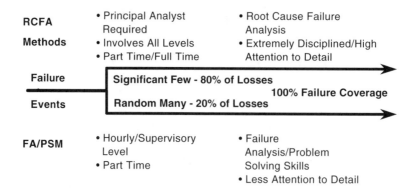

FIGURE 6.1 The two-track approach to failure avoidance.

Conversely, when we look at what data collection strategy would be employed on a sporadic event, we find that frequency does *not* work in our favor. Under these circumstances, our detective may be investigating a single homicide and be relying on the evidence at that scene only. This would mean that we must be very diligent about collecting the data from that scene before it is tampered with. Related to our environments, when a sporadic event occurs, we must be diligent at that time to collect the data in spite of the massive efforts to get production running again.

PRESERVING EVENT DATA

The first step in the PROACT® process, as is the case in any investigative or analytical process, is to preserve and collect pertinent data. Before we discuss the specifics of how to collect various forms of data and when to collect it, let's take a look at the psychological side of why people should assist in collecting data from an event scene.

Let's create a scenario: we are a mechanic in a manufacturing plant. We just completed a 10-day shutdown of the facility to perform scheduled maintenance. Everyone knows at this facility that when the plant manager says the shutdown will last 10 days and no more, we do not want to be the one responsible for extending it past 10 days. A situation arises in the ninth day of the shutdown where during an internal preventive maintenance inspection, we find that a part has failed and must be replaced. In good faith we request the part from the storeroom. The storeroom personnel inform us that the particular part is stocked out and it will take four weeks to expedite the order from the vendor. Knowing this is the 9th day of the 10-day shutdown, we decide to make a "band aid" repair because we do not want to be the person to extend the shutdown. We rationalize that the band aid will hold for the four-week duration as we have gotten away with it in the past. So we install a not "like-for-like" part in preparation for the start-up of the process.

Within 24 hours of start-up, the process fails catastrophically and all indications lead to the area where the band aid was installed. A formal RCA team is amassed and they assign us to collect some parts data from the scene immediately. Given the witch-hunting culture that we know exists, why should we uncover data/evidence that will incriminate us? While this is a hypothetical scenario, it could very well

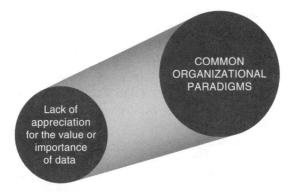

FIGURE 6.2 Typical reasons why event data is not collected.

represent many situations in any industry. What is the incentive to collect event data? After all, this is a time-consuming task, it will lead to people who used poor judgment, and therefore, management could witch-hunt.

These are all very valid concerns. We have seen the good, the bad, and the ugly created by these concerns. The fact of the matter is that if we wish to uncover the truth, the real root causes, we cannot do so without the necessary data. Think about any investigative or analytical profession; the first step is always to collect data. Is an detective expected to solve a crime without any evidence or leads? Is an NTSB investigator expected to solve the reasons for an airline crash without any evidence from the scene? Do doctors make diagnoses without any information as to symptoms? If these professionals see the necessity of data and information to draw conclusions, then we certainly must recognize the correlation to RCA.

Based on our experience, we have seen a general resistance to data collection for RCA purposes. We can draw two general conclusions from our experience (see also Figure 6.2):

1. People are resistant to collecting event data because they do not appreciate the value of the data to an analysis or analyst.
2. People are resistant to collecting data because of the paradigms that exist with regards to witch hunting and managerial expectations.

The first conclusion we see is the minor one of the two. Oftentimes production of any facility is the ruling body. After all, we are paid to produce quality product whether that product is oil, steel, package delivery, or quality patient care. When this mentality is dominant, it forces us to react with certain behaviors. If production is paramount, then whenever an event occurs, we must clean it up and get production started as quickly as possible. The focus is not on why the event occurred, rather it is on the fact that it did occur and we must get back on line.

This paradigm can be overcome merely with awareness and education. Management must first commit to supporting RCA both verbally and on paper. We discussed earlier in the management support chapter that demonstrated actions are what are seen as "walking the talk" and one of those actions was issuing an RCA policy or procedure. This would make data collection a requirement instead of an option. Second, it is not just enough to support it, but we must link with the individuals who must physically collect the data. They must clearly understand *why* they should collect the data and *how* to do it properly.

We should link with people's value systems and show them the purpose of data collection. If we are an operator in a steel mill and the first one to an event scene, we should understand what is important information vs. unimportant to an RCA. For instance, we can view a broken shaft as an item to clean up or as an integral piece of information to a metallurgist. If we understand how important the data we collect is to an analysis, we will see and appreciate why it should be collected. If we do not understand or appreciate its value, then the task is seen as a burden to our already full plate. Providing everyone with basic training in proper data collection procedures can prove invaluable to any organization.

We have seen of late the potential consequences of poor data collection efforts in some high-profile court cases. Allegations are made as to the sloppy handling of evidence in lab work, improper testing procedures, improper labeling, and contaminated samples. Issues like these can lose your case.

Providing the above support and training overcomes one hurdle. But it does not clear the hurdle of perceived witch hunting by an organization. Many people will choose not to collect data for fear that they may be targeted based on the conclusion drawn from the data. This is a very prominent cultural issue that must be addressed in order to progress with RCA. We cannot determine root causes if a witch-hunting culture is prevalent.

THE ERROR-CHANGE PHENOMENON

Research indicates that there is an average number of errors that must queue up in a particular pattern for a catastrophic event to occur. The error chain concept,* "...describes human error accidents as the result of a sequence of events that culminate in mishaps. There is seldom an overpowering cause, but rather a number of contributing factors of errors, hence the term *error chain*. Breaking any one link in the chain might potentially break the entire error chain and prevent a mishap." This research comes from the aviation industry and is based on the investigation of more than 30 accidents or incidents. This has been our experience as well in investigating industrial failures.

Flight Safety International states that the fewest links discovered in any one accident was four, the average being seven.** Our experience in industrial applications shows the average number of errors that must queue up to be between 10 and 14. To us, this is the core to understanding what an analyst needs in order to understand why undesirable events occur.

We like referring to it as error-change relationships. First we must define some terms in order to communicate more effectively. We will use a modification of James Reasons'*** definition of error for our RCA purposes. An error to us is defined as

* Flight Safety International, Crew Resource Management Workshop, September 1993.
** Flight Safety International, Crew Resource Management Workshop, September 1993.
*** Reasons, James. *Human Error.* New York: Cambridge University Press, 1990.

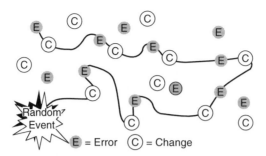

FIGURE 6.3 The error-change phenomenon.

an action planned, but not executed according to the plan. This means that we intended on a satisfactory outcome and it did not occur. We, in some manner, deviated from our intended path. The change, as a result of an error in our environment, is something that is perceptible to the human senses. An example might be that we commit an error by misaligning a shaft; the change will be that an excessive vibration occurs as a result. A nurse administering the wrong medication to a patient is an error; the adverse reaction is the perceptible change. These series of errors and associated changes are occurring around us every day (see Figure 6.3). When they queue up in a particular pattern, catastrophic occurrences happen.

Knowing this information, we would like to make two points:

1. We as human beings have the ability through our senses to be more aware of our environments. If we sharpen our senses, we can detect these changes and take action to prevent the error chain from running its course. Many of our organizational systems are put in place to recognize these changes. For example, the predictive maintenance group's sole purpose is to use testing equipment to identify changes within the process and equipment. If changes are not within acceptable limits, actions are taken to make them within acceptable limits.
2. By witch hunting the last person associated with an event, we give up the right to the information that person possesses on the other errors that lead up to the event. If we discipline a person associated with the event because our culture requires a head to roll, then that person (or anyone around them) will not be honest about why they made decisions that resulted in errors.

In a later chapter on analyzing the data (Chapter 8), we will explore what we call a *logic tree,* which is a graphical representation of an error-change chain based on this research. We discuss this research at this point because it is necessary to understand that any investigation or analysis cannot be performed without data. We have enough experience in the field application of RCA to make a general statement that the physical activity of obtaining such data can have many organizational barriers. Once these barriers are recognized and overcome, the task of actually preserving and collecting the data begins.

THE 5-Ps CONCEPT

PReserving failure data is the "PR" in PROACT®. In a typical high-profile RCA, an immense amount of data is collected and then must be organized and managed. As we go through this discussion, we will relate how to manage this process manually vs. with software. We will discuss automating your RCA in Chapter 11.

Consider this scenario: a major upset just occurred in our facility. We are charged with collecting the necessary data for an investigation. What is the necessary information to collect for an investigation or analysis? We use a 5-Ps approach, where the Ps stand for the following:

1. **P**arts
2. **P**osition
3. **P**eople
4. **P**aper
5. **P**aradigms

Virtually anything that needs to be collected from an event scene can be categorized under one of these headings. Many items will have shades of gray and fit under two headings, but the important thing is to capture the information and slot it under one heading. This categorization process will help manage the data for the analysis.

Let's use the parallel of the detective again. What do we see detectives and police routinely do at a crime scene? We see the police rope off the area to preserve the positional information. We see detectives interviewing people that may be eyewitnesses. We see forensic teams "bagging and tagging" evidence or parts. We see a hunt begin for information or a paper trail of a suspect that may involve past arrests, insurance information, financial situation, etc. And last, as a result of the interviews with the observers, we draw tentative conclusions about the situation such as, "…he was always at home during the day and away at night. We would see children constantly visiting for five minutes at a time. We think he is a drug dealer." These are the paradigms that people have about situations that are important because if they believe these paradigms, then they are basing decisions on them. This can be dangerous.

PARTS

Parts will generally mean something physical or tangible. The potential list is endless depending on the industry where the RCA is conducted. For a rough sampling of what is meant by parts, please review the following lists:

Continuous process industries — (oil, steel, aluminum, paper, chemicals, etc.):

- Bearings
- Seals
- Couplings
- Impellers

- Bolts
- Flanges
- Grease samples
- Product samples
- Water samples
- Tools
- Testing equipment
- Instrumentation
- Tanks
- Compressors
- Motors

Discrete product industries — (automobiles, package delivery, bottling lines, etc.):

- Product samples
- Conveyor rollers
- Pumps
- Motors
- Instrumentation
- Processing equipment

Healthcare — (hospitals, nursing homes, out patient care center, etc.):

- Medical diagnostic equipment
- Surgical tools
- Gauze
- Fluid samples
- Blood samples
- Biopsies
- Medicines
- Syringes
- Testing equipment

This is just a sampling to give a feel for the type of information that is considered *parts*.

POSITION

Position data is the least understood and what we consider to be the most important. Positional data comes in the form of two different dimensions, one being physical space and the second being point in time. Positions in terms of space are vitally important to an analysis because of the facts that can be deduced.

When the space shuttle Challenger exploded on January 28, 1986, it was approximately 5 miles in the air. Films from the ground provided millisecond by millisecond footage of the parts that were being dispersed from the initial cloud. From this positional information, trajectory information was calculated and search and recovery groups were assigned to approximate locations of where vital parts were located. Approximately

93,000 square miles of ocean were involved in the search and recovery of shuttle evidence in the government investigation.* While this is an extreme case, it shows how position information is used to determine, among other things, force.

While on the subject of the shuttle Challenger, other positional information that should be considered is why it was the right solid rocket booster (SRB) and not the left; why it was the aft (lower) field joint attachment vs. the upper field joint attachment; and why the leak was at the O-ring on the inside diameter of the SRB vs. the outside diameter. These are questions regarding positional information that had to be answered.

Now let's take a look at positions in time and their relative importance. Monitoring positions in time at which undesirable outcomes occur can provide information for correlation analysis. By recording historical occurrences we can plot trends that identify the presence of certain variables when these occurrences happen. Let's take a look at the shuttle again. Most of us remember the incident and the conclusion reported to the public: an O-ring failure resulted in a leak of solid rocket fuel. If we look at the positional information from the standpoint of time, we would learn that the O-rings had evidence of secondary O-ring erosion on 15 of the previous 25 shuttle launches.** When the SRBs are released they are parachuted into the ocean, retrieved, and analyzed for damage. The correlation of these past launches, which incurred secondary O-ring erosion, showed that low temperatures were a variable. The position in time information aided in this correlation.

Now moving into more familiar environments, we can review some general or common positional information to be collected at most any organization:

- Physical position of parts at scene
- Point in time of current and past occurrences
- Position of instrument readings
- Position of personnel at time of occurrence(s)
- Position of occurrence in relation to overall facility
- Environmental information related to position of occurrence such as temperature, humidity, wind velocity, etc.

Again, this is just a sampling to give individuals the concept of what we mean by positional information.

Figure 6.4 represents a sample of a graphic used to mark positions of various temperature distributions within a boiler with failed tubes.

PEOPLE

The *people* category is the more easily defined "P." This is simply who we need to talk to initially in order to obtain information about an event. The people we must talk to first should typically be the physical observers or witnesses to the event. Efforts to obtain such interviews should be relentless and immediate. We risk the chance of losing direct observation when we interview days after an event occurs.

---

* *Challenger: Disaster and Investigation.* Cananta Communications Corp.1987.
** Lewis, Richard S. *Challenger: The Final Voyage.* New York: Columbia University Press, 1988.

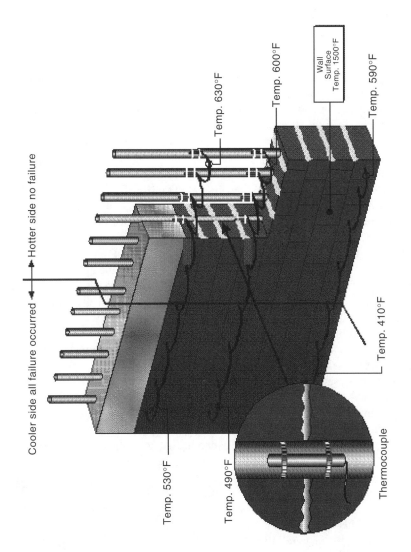

FIGURE 6.4 Mapping example of sulfur burner boiler.

We will ultimately lose some degree of short-term memory recollection and also risk the observers having talked to others about their opinion of what happened. Once an observer discusses such an event with another outsider, they will tend to reshape their direct observation with the new perspectives.

We have always regarded the goal of an interview with an observer to be that we must be able to see through their eyes what they saw at the scene. The description must be that vivid and it is up to the interviewer to obtain that clarity through the questioning.

Interviewing skills are necessary in such analytical work. People must feel comfortable around an interviewer and not intimidated. A poor interviewing style can ruin an interview and subsequently an analysis or investigation. A good interviewer will understand the importance and value of body language. Experts estimate that approximately 55 to 60% of all communication between people is through body language. Approximately 30% is through the tonal voice and 10 to 15% is through the spoken word.* This is very important when interviewing because it emphasizes the need to interview in person rather than over the telephone. Lawyers are professionals at reading the body language of their clients, their opposition, and the witnesses. Body language clues will direct their next move. This should be the same for interviews associated with an undesirable outcome. The body language will tell the interviewer when they are getting close to information they desire, and this will direct the line and tone of subsequent questioning.

When in situations of interviewing during the course of an RCA, it is also important to consider the logistics of the interview. Where is the appropriate place to interview? How many people should we interview at a time? What types of people should be in the room at the same time? How will I record all the information? Preparation and environment are very important factors to consider.

We discussed the interviewing environment and the ideal number of people in an interview in Chapter 4. Those same pointers will hold true when interviewing for the actual root cause analysis (RCA) vs. the modified FMEA.

We have most success in interviews when the interviewees are from various departments, and more specifically from different "kingdoms." We will define a kingdom as an entity that builds its castles within facilities and tends not to communicate with others. Examples can be maintenance vs. operations, labor vs. management, doctors vs. nurses, hourly vs. engineers, etc. When such groups get together, they learn a great deal about the other's perspective and tend to earn a respect for each other's position. An added benefit of an RCA is that people actually start to meet and communicate with each other from different levels and areas.

In an ideal situation, an associate analyst can take the notes while the interviewer focuses on the interview. It is not recommended that recording devices be used in routine interviews as they are intimidating and people believe that the information may be used against them at a later date. In some instances where severe legal liabilities may be at play, legal counsel may impose such actions. However, if they do, they are generally doing the interviewing. In the case of most chronic failures or events, such extremes are rare.

---

* Lyle, Jane. *Body Language*. London: The Hamlyn Publishing Group Limited, 1990.

Common people to interview will again be based on the nature of the industry and the event being analyzed. As a sample of potential interviewees, please consider the following list:

- Observers
- Maintenance personnel
- Operations personnel
- Management personnel
- Administration personnel
- Clinicians
- Technical personnel
- Purchasing personnel
- Storeroom personnel
- Vendors representatives
- Original equipment manufacturers (OEM)
- Other similar sites with similar processes
- Inspection/quality control personnel
- Safety personnel
- Environmental personnel
- Lab personnel
- Outside experts

As stated previously, this list is just to give a feel for the variety of people that may provide information about any given event.

PAPER

Paper data is probably the most understood form of data. Being in an information age where we have instant access to data through our communications systems, we tend to be able to amass a great deal of paper data. However, we must make sure that we are not collecting paper data for the sake of developing a big file. Some companies we have seen seem to feel they are getting paid based on the width of the file folder. We must make sure that the data we are collecting is relevant to the analysis at hand.

Keep in mind our detective scenarios and the fact that they are always preparing a solid case for court. Paper data is one of the most effective pieces of evidence in court. Solid, organized documentation is the key to a winning strategy.

Typical paper data examples are:

- Chemistry lab reports
- Metallurgical lab reports
- Specifications
- Procedures
- Policies
- Financial reports
- Training records

- Purchasing requisitions/authorizations
- Nondestructive testing results
- Quality control reports
- Employee file information
- Maintenance histories
- Production histories
- Medical histories
- Safety records information
- Internal memos/e-mails
- Sales contact information
- Process & instrumentation drawings
- Medical charts
- Labeling of equipment/products
- Distributive control system (DCS) strips
- Statistical process control/statistical quality control information (SPC/SQC)

In Chapter 11, "Automating Root Cause Analysis," we will discuss how to keep all this information organized and properly documented in an efficient and effective manner.

PARADIGMS

Paradigms have been discussed throughout this text as a necessary foundation for understanding how our thought processes affect our problem-solving abilities. But exactly what are paradigms? We will base the definition we use in RCA on Futurist Joel Barker's definition as follows:

> A paradigm is a set of rules and regulations that: (1) Defines boundaries; and (2) tells you what to do to be successful within those boundaries. (Success is measured by the problems you solve using these rules and regulations.)*

This is basically how each individual views the world and reacts and responds to situations that arise. This inherently affects how we approach solving problems and will ultimately be responsible for our success or failure in the RCA effort.

Paradigms are a by-product of interviews carried out in this process and discussed earlier in this chapter. Paradigms are recognizable because repetitive themes are expressed in these interviews from various individuals. How an individual sees the world is a mindset. When a certain population shares the same mindset, it becomes a paradigm. Paradigms are important because even if they are false, they represent the beliefs on which we base our decision making. Therefore, true paradigms represent reality to the people that possess them.

---

* Barker, Joel. *Discovering the Future: The Business of Paradigms*. Elmo, MN: ILI Press, 1989.

Below is a list of common paradigms we see in our travels. We are not making a judgment as to whether they are true or not, but rather that they affect judgment in decision making.

- We do not have time to perform RCA.
- We say safety is number one, but when it comes down to brass tacks, cost is really number one.
- This is impossible to solve.
- We have tried to solve this for 20 years.
- It is old equipment, it is supposed to fail.
- We know because we have been here for 25 years.
- This is another program-of-the-month.
- We do not need data to support RCA because we know the answer.
- This is another way for management to "witch hunt."
- Failure happens; the best we can do is sharpen our response.
- RCA will eliminate maintenance jobs.

Many of these statements may sound familiar. But think about how each statement could affect problem-solving abilities. Consider these if-then statements.

- If we see RCA as another burden (and not a tool), then we will not give it a high priority.
- If we believe that management values profit more than safety, then we may rationalize at some time that bending the safety rules is really what our management wants us to do.
- If we believe that something is "impossible" to solve, then we will not solve it.
- If we believe that we have not been able to solve the problem in the past, then no one will be able to solve it.
- If we believe that equipment will fail because it is old, then we will be better prepared to replace it.
- If we believe RCA is the program-of-the-month, then we will wait it out until the fad goes away.
- If we do not believe data collection is important, then we will rely on word of mouth and allow ignorance and assumption to penetrate an RCA as fact.
- If we believe that RCA is a witch-hunting tool, then we will not participate.
- If we believe failure is inevitable, then the best we can do is become a better responder.
- If we believe that RCA will eventually eliminate our job, then we will not let it succeed.

Our purpose with these if-then statements is to show the effect that paradigms have on human decision making. When human errors in decision making occur, they are the triggering mechanisms for a series of other subsequent errors until the undesirable event surfaces and is recognized.

TABLE 6.1
Data Fragility Rankings

5-Ps	Fragility Ranking
Parts	1
Position	2
Paper	3
People	1
Paradigms	4

We have discussed in detail the error-change phenomenon and the 5-Ps. Now we must discuss how we get all of this information. When an RCA has been commissioned, a group of data collectors must be assembled to brainstorm what data will be necessary to start the analysis. This first session is just that, a brainstorming session of data needs. This is not a session to analyze anything. The group must be focused on data needs and not be distracted by the premature search for solutions. The goal of this first session should not be to collect 100% of the data needed, rather attempts should be made to be in the 60 to 70% range. All of the obvious surface data should be collected first and also the most fragile data. Table 6.1 describes the normal fragility of data at an event scene. By fragility we mean the prioritization of the 5-Ps in terms of which is most important to collect first, second, third, and so on.

You will notice that people and parts are tied for first. This is not an accident. As we discussed earlier in this chapter, the need to interview observers is immediate in order to obtain direct observation. Positional information is equally important because it is the most likely to be disturbed the quickest. Therefore, attempts to get such data should be performed the soonest. Parts are second because if there is not a plan to obtain them, then they will typically end up in the trash can. Paper data is generally static with the exception of process or online production data (DCS, SPC/SQC). Such new technologies allow for automatic averaging of data to the point that if the information is not retrieved within a certain time frame, then it can be lost forever. Paradigms are last because we wish we could change them faster, but modifying behavior and belief systems takes time.

One preparatory step for analysts should be to have a data collection bag always prepared. Many times such events occur when we least expect it. We do not want to have to be running around collecting a camera, plastic bags, etc. If it is all in one place, it is much easier to go prepared in a minute's notice. Usually go models are from other emergency response occupations such as doctors' bags, fire departments, police departments, etc. They always have most of what they need accessible at any time. Such a bag (in general) may have the following items:

- Caution tape
- Masking tape
- Plastic bags

Data Type: _____   Responsible: _____

Completion Date:__/__/__

Person to Interview/Item to be Collected: _____
_____
_____

Data Collection Strategy: _____
_____
_____
_____

FIGURE 6.5 Sample 5-Ps data collection form.

- Gloves
- Safety glasses
- Ear plugs
- Adhesive labels
- Marking pens
- 35mm camera w/film
- Video camera (if possible)
- Marking paint
- Tweezers
- Pad and pen
- Ruler
- Sample vials
- Wire tags to ID equip

This is of course a partial listing and depends on the organization and nature of work; other items would be added or deleted from the list.

Figure 6.5 shows a typical data collection form used for manually organizing data collection strategies for an RCA team. It was designed to collect the following information:

1. **Data Type** — List which of the 5-Ps this form is directed at. Each "P" should have its own form.
2. **Responsible** — The person responsible for making sure the data is collected by the assigned date.
3. **Completion Date** — During brainstorming session list all data necessary to collect for each "P" and assign a date by which the data should be collected.
4. **Data Collection Strategy** — This space is for actually listing the plan of how to obtain the previously identified data to collect.
5. **Date to Be Collected by** — Date by which the data is to be collected and ready to be reported to team.

Figure 6.6 is a completed sample data collection worksheet.

Data Type: Paper_____ Responsible: John Smith_____

Completion Date: 02 / 02 / 97

Person to Interview/Item to be Collected: Shift Logs_____

_____
_____

Data Collection Strategy:___Have shift foreman collect the
shift logs when pump 235 fails and deliver it to John Smith
within 1 day._____

_____

FIGURE 6.6 Completed data collection form.

7 Ordering the Analysis Team

When a sporadic event typically occurs in any organization, an immediate effort is made to form a task team to investigate why such an undesirable event occurred. What is the typical make-up of such a task team? We see the natural tendency of managements to assign the cream-of-the-crop experts to both lead and participate on such a team. While well-intended, there are some potential disadvantages to this thought process.

Let's paint a scenario in a manufacturing setting (even though it could happen anywhere else). A sulfur burner boiler fails due to tube ruptures. The event considerably impacts production capabilities when it occurs. Maintenance histories confirm that such an occurrence is chronic as it has happened at least once a year for the past ten years. Therefore, mean time between failure (MTBF) is approximately one year. This event is a high priority in the mind of the plant manager as it is impacting the facility's ability to meet corporate production goals. The manager is anxious for the problem to go away and makes the logical deduction that if tubes are rupturing in this boiler, then it must be a metals issue. Based on this premise, I naturally would want to have my best people on the team. I would assign my top metallurgist as the team leader because he has been with the company the longest and has the most experience in the department. On the team I will provide him the resources of his immediate staff to dedicate the time to solve the problem. Does this sound familiar? The logic appears sound. Why wouldn't this strategy work?

Let's review what typically happens next. I have a team of, say, five metallurgists. They are brainstorming all the reasons these tubes could be bursting. At the end of their sessions, they conclude that more exotic metals are required and that the tube materials should be changed to endure the atmosphere in which they operate. Problem solved! However, this is the same scenario that went on for ten years, and we kept replacing the tubes year after year and they still kept rupturing.

Think about what just went on with that team. Remember our earlier discussion about paradigms and how people view the world. How do we think the team of metallurgists view the world? They all share the same "box." They have similar educational backgrounds, similar experiences, similar successes, and similar training. That is what they know best: metallurgy. Any time we put five metallurgists on a team we will typically have a metallurgical answer. The same goes for expertise in any discipline. This is the danger of not having technical diversity on a team and also of letting an expert lead a team on an event in which he is the expert.

The end to the story above is that eventually an engineer of a different discipline was assigned as the leader of the team. The new team had metallurgists as well as mechanical and process engineers. The end result of the thorough RCA was that the tubes were in a specific location of the boiler that was below the dew point for sulfuric acid and therefore the tubes were corroding due to the environment. The solution was to return to the base metals and move the tubes 18" forward where the temperature was acceptable. When team leaders are not experts, they can ask any question they wish of the team members who may be experts. However, this luxury is not afforded to experts who lead teams because they are generally perceived by team members as knowing all, therefore they cannot ask the perceived obvious or "stupid" question. While this seems a trivial point, it can, in fact, be a major barrier to success.

To avoid this trap of narrow-minded thinking, let's explore the anatomy of an ideal RCA team. The purpose of a diverse team is to provide synergism where the whole is greater than the sum of the parts. Anyone who has participated in the survival teaming games and outings will agree that when different people of different backgrounds come together for a team purpose, their outcomes are better as a team than if they had pursued the problem as an individual.

Teams have long been a part of the quality era and are now commonplace in organizations. Working in a team can be the most difficult part of our work environments because we will be working with others who may not agree with our views. This is the reason that teams work: people disagree. When people disagree, each side must make a case to the other why their perspective is correct. To support this view, a factual basis must be provided rather than conventional wisdom. This is where the learning comes in and teams progress. We always use the line that "if a team is moving along in perfect harmony, then changes need to be made in team make-up." We must seek debate and conflict in order to progress. While this may seem difficult to deal with, it will promote success of the team's charter.

What is a team?

> A team is a small number of people with complementary skills who are committed to a common purpose, performance goals, and approach for which they hold themselves mutually accountable.*

A team is different from a group. A group can give the appearance of a team; however, the members act individually rather than in unison with others.

Let's now explore the following key elements of an ideal RCA team structure:

- Team member roles and responsibilities
- Principal analyst characteristics
- The challenges of RCA facilitation
- Promote listening skills
- Team codes of conduct
- Team charter
- Team critical success factors
- Team meeting schedules

---

* Katzenbach, Jon R. & Smith, Douglas K. *The Wisdom of Teams.* Boston, Massachusetts: Harvard Business School Press, 1994.

TEAM MEMBER ROLES AND RESPONSIBILITIES

Many views about ideal team size are prevalent. The situation that created the team will generally define how many members are appropriate. However, from an average standpoint for RCA, it has been our experience that between four and seven members are ideal and beyond ten are too many. Having too many people on a team can force the goals to be prolonged due to the dragging on of too many opinions.

Who are the core members of an RCA team? They are:

- The principal analyst (PA)
- The associate analyst
- The experts
- Vendors
- Critics

THE PRINCIPAL ANALYST (PA)

Each RCA team needs a leader. This is the person who will ultimately be held accountable by management for results. The principal anlyst is the person who will drive success and accept nothing less. This leader will either make or break the team.

This person is responsible for the administration of the team efforts, the facilitation of the team members according to the PROACT® philosophy, and the communication of goals and objectives to management oversight personnel.

THE ASSOCIATE ANALYST

This position is often seen as optional; however, if the resources are available to fill it, it is of great value. The associate analyst is basically the "leg man" for the PA. This person will execute many of the administrative responsibilities of the PA such as inputting data, issuing meeting minutes, arranging for conference facilities, arranging for audio/visual equipment, obtaining paper data such as records, etc. This person relieves much of the administration from the PA, allowing the PA more time to focus on team progress.

THE EXPERTS

This is basically the core make-up of the team. These are the individuals that the PA will facilitate. They are the "nuts and bolts" experts on the issue being analyzed. These individuals will be chosen based on their backgrounds in relation to the issue being analyzed. For instance, if we are analyzing an equipment breakdown in a plant, we may choose to have operations, maintenance, and engineering personnel represented on the team. If we are exploring an undesirable outcome in a hospital setting, we may wish to have doctors, nurses, lab personnel, and quality/risk management personnel on the team. In order to develop accurate hypotheses, experts are absolutely necessary on the team. Experts will aid the team in generating hypotheses and also verifying them in the field.

VENDORS

Vendors are an excellent source of information about their products. However, in our opinion, they should not lead such an analysis when their products are involved in an event. Under such circumstances, we want the conclusions drawn by the team to be unbiased so that they have credibility. It is often very difficult for a vendor to be unbiased about how its product performed in the field. For this reason, we suggest that vendors participate on the team, but not lead the team.

Vendors are great sources of information for generating hypotheses about how their products could not perform to expectations. However, they should not be permitted to prove or disprove their own hypotheses. We often see the vendor blame nonperformance on the way in which the product was handled or maintained. It is something that the customer did rather than a flaw in the product. We are not saying that the customer is always right, but from an unbiased standpoint, we must explore both the possibilities that the product has a problem as well as that the customer could have done something wrong. Remember, facts lead such analyses, not assumptions!

CRITICS

We have never come across a situation in our careers where we had difficulty in locating critics. Everyone knows who he or she is in the organization. However, sometimes they get a bad reputation just because they are curious. Critics are typically people who just do not see the world the way that everyone else does. They are really the "devil's advocates." They will force the team to see the other side of the tracks and find holes in logic by asking persistent questions. They are often viewed as uncooperative and not team players. But they are a necessity to a team.

Critics come in two forms: (1) constructive and (2) destructive. Constructive critics are essential to success and are naturally inquisitive individuals who take nothing (or very little) at face value. Destructive critics stifle team progress and are more interested in overtime and donuts than in successfully obtaining the team charter.

PRINCIPAL ANALYST CHARACTERISTICS

The PA typically has a hard row to hoe. If RCA is not part of the culture, then the PA is going against the grain of the organization. This can be very difficult to deal with if we are people who have difficulty in dealing with barriers to success. Over the years we have noted the personality traits that make certain PAs stars whereas others have not progressed. Below is a list of the key traits that our most successful analysts possess (many of them lead the analyses that are listed in the case histories of this text):

- *Unbiased* — While we discussed this issue earlier, this is a key trait to the success of any RCA. The leader of an RCA should have nothing to lose and nothing to gain by the outcome of the RCA. This ensures that the outcomes are untainted and credible.

- *Persistent* — Individuals who are successful as PAs are those who do not give up in times of adversity. They do not retreat at the first sign of resistance. When they see roadblocks they immediately plan to go through them or around them. "No" is not an acceptable answer. "Impossible" is not in their vocabulary. They are painstakingly persistent and tenacious.
- *Organized* — PAs are required to maintain the organizational process of the RCA. They are responsible for organizing all the information being collected by the team members and putting it into an acceptable format for presentation. Such skills are extremely helpful in RCA.
- *Diplomatic* — Undoubtedly PAs will encounter situations where upper level management or lower level individuals will not cooperate. Whether it is maintenance not cooperating with operations, unions boycotting teaming, administration not willing to provide information, or doctors not willing to participate on teams due to legal counsel advice, political situations will arise. A great PA will know how to handle such situations with diplomacy, tactfulness, and candidness. The overall objective in all these situations is to get what we want. We work backwards from that point in determining the means to attain the end.

THE CHALLENGES OF RCA FACILITATION

For all of us who have ever facilitated any type of team, we can surely appreciate the need to possess the characteristics described above. We can also appreciate the experience that such tasks have provided us about dealing with the human being. Below we explore common challenges faced when facilitating a typical RCA team.

BYPASSING THE RCA DISCIPLINE AND GOING STRAIGHT TO SOLUTION

As we all have experienced in our daily routines, the pressure of the daily production overshadows our intentions of doing things right and stepping back and looking at the big picture. This phenomenon becomes apparent when we organize an RCA team that is well versed in how to repair things quickly to get production up and running now. Such teams will be inclined to pressure the RCA facilitator to hurry up and implement their solution(s).

FLOUNDERING OF TEAM MEMBERS

One of the more predominant problems with most RCA attempts is lack of discipline and direction of method. This results in the team becoming frustrated because it appears that the team is going around in circles and getting nothing done. Also, if there are team members who have not been educated in the RCA methodology being employed, then they can see no light at the end of the tunnel. Such team members tend to get bored quickly and lose interest.

ACCEPTANCE OF OPINIONS AS FACTS

This often occurs using methodologies that promote solutions before proving that hypotheses are factual. We have all seen where we are so pressured to get back to normal or the status quo, that we tend to accept people's opinions as facts so that we can come to consensus quickly and try to implement solutions. Often this haste results in spending money that does not solve the problem and is akin to the trial-and-error approach.

DOMINATING TEAM MEMBERS

Generally most teams that are organized under any circumstances have one strong-willed person who tends to impose his or her personality on the rest of the team members. This can result in the other team members being intimidated and not participating, but more likely it pressures accepting opinion as fact (described above).

RELUCTANT TEAM MEMBERS

We have all participated on teams where some members were much more introverted than others. It is not that they do not have the experience or talent to contribute, but their personality is simply not an outgoing one. Sometimes people are reluctant to participate because they feel that authority is in the room and they do not want to appear as not asking the right questions, so they say nothing at all and do not rock the boat.

GOING OFF ON TANGENTS

Again, these characteristics can (and do) appear on any team. They are functions of team dynamics that happen when humans work together. An RCA facilitator is charged with sticking to the discipline of the RCA method. This includes keeping the team on track and not letting the focus drift.

ARGUING AMONG TEAM MEMBERS

Nothing can be more detrimental to a team than its members engaging in destructive arguments due to close-mindedness. There is a clear difference between argument and debate. Arguments tend to get polarized and each side takes a stance and will not budge. The goal of an argument in these cases is for the other side to agree with you totally, not to come to consensus. Debate promotes consensus, which requires a willingness to meet in the middle if necessary.

PROMOTE LISTENING SKILLS

Obviously, most of the team dynamics issues that we are discussing are not just pertinent to RCA, but to any team. While the concept of listening seems simplistic, most of us are not adept at its use.

Many of us always state that we are not good at remembering names. If we look back at a major cause of this, it is because we never actually listen to people when they introduce themselves. Most of the time when someone is introducing themselves to us, we are more preoccupied with preparing our response than actually listening to what they are saying.

Next time you meet people, concentrate on remembering to actually listen to their introductions and taking a snapshot memory of their faces with your eyes. You will be amazed at how that impression will log into your long-term memory and pop up the next time that you see them.

The following are listening techniques that may be helpful when we organize our RCA teams.

- One person speaks at a time — This may appear to be common sense and a mere matter of respect, but how often do we see this rule broken? We obviously cannot be listening if there is input from more than one person at a time.
- Do not interrupt — Interrupting is rude; let people finish their point while you listen. We will have plenty of time to formulate an educated response. Sometimes we think that if we are the fastest and the loudest to make statements then we will gain ground. We can all watch the Springer Show and know that is not the case.
- React to ideas not people — This is a very important point and should not be forgotten. Even if we disagree with other team members, *never* make it a personal issue. We may disagree with someone's ideas, but that does not mean that it is a personal issue between them and us. This is totally unproductive and will cause digression rather than progression if permitted to exist.
- Separate facts from conventional wisdom — Just like in the courtroom, in our debates, we must separate facts from conventional wisdom. After all, in RCA, the entire discipline is based on facts. Conventional wisdom originates from opinions, and if not proven, will result in assumptions.

TEAM CODES OF CONDUCT

Codes of conduct were most popular within the quality circles and the push for teaming. They vary from company to company, but what they all have in common is the desire to make meetings more efficient and effective. Codes of conduct are merely sets of guidelines by which a team agrees to operate. Such codes are designed to enhance the productivity of meetings. The following are a few examples of common sense codes of conduct:

- All members will be on time for scheduled meetings.
- All meetings will have an agenda that is to be followed.
- Everyone's ideas will be heard.
- Only one person will speak at a time.
- A three knock rule, where a person politely knocks on the table, is used to provide an audio indicator that the speaker is going off track of the topic being discussed.
- A holding area is provided on the easel pad where topics are placed for consideration on the next meeting agenda because they are not appropriate for the meeting at hand.

This is just a sampling to give an idea as to what team meeting guidelines can be like. Many of our clients who have embraced the quality philosophy will have such codes of conduct framed and posted in all conference rooms. This provides a visual reminder that will encourage people to abide by such guidelines in an effort not to waste people's time.

TEAM CHARTER

The team's charter (sometimes referred to as a mission) is a one-paragraph statement delineating why the team was formed in the first place. This statement shall serve as the focal point for the team. Such a statement should be agreed upon not only by the team, but also by the managers overseeing the team's activities. This will align everyone's expectations as to direction and results.

The following is a sample team charter reflecting a team that was organized to analyze a mechanical failure:

> To identify the root causes of the ongoing motor failures occurring on pump CP-220, which includes identifying deficiencies in, or lack of, organizational systems. Appropriate recommendations for root causes will be communicated to management for rapid resolution.

TEAM CRITICAL SUCCESS FACTORS (CSFs)

Critical success factors are guidelines by which we will know that we are successful. I have heard CSFs also referred to as key performance indicators (KPIs) and other nomenclature. Regardless of what we call them, we should set some parameters that will define the success of the RCA team's efforts. This should not be an effort in futility to list a hundred different items. We recommend no more than eight should be designated per analysis. Experience will support that typically many are used over and over again on various RCA teams. The following are a few samples of CSFs:

- A disciplined RCA approach will be used and adhered to.
- A cross-functional section of plant personnel/experts will participate in the analysis.
- All analysis hypotheses will be verified or disproved with factual data.
- Management agrees to fairly evaluate the analysis team's findings and recommendations upon completion of the RCA.
- No one will be disciplined for honest mistakes.
- A measurement process will be used to track the progress of implemented recommendations.

TEAM MEETING SCHEDULES

We are often asked, "What is the average time of an analysis?" The answer is how important is the resolution of the event. The higher the priority of the event being analyzed, the quicker the process will move. We have seen high priority given to

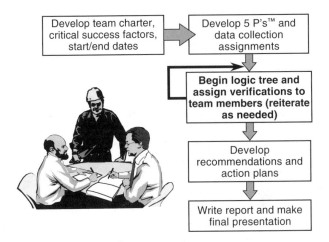

FIGURE 7.1 RCA team process flowchart.

events to the degree that full-time teams are assigned and resources and funds are unlimited to find the causes. These are usually situations where there must be a visual demonstration of commitment on behalf of the company because the nature of the event had been picked up by the media and now the public wants an answer. These are usually analyses of sporadic vs. chronic events. The space shuttle Challenger disaster is such an example where the public's desire to know forced an unrelenting commitment on behalf of the government.

Unfortunately, such attention is rarely given to events that do not hurt individuals, do not destroy equipment, and do not require analysis due to regulatory compliance. These are usually indicative of chronic events.

As we tell our students, we provide the architecture of an RCA methodology. It will not work the same in every organization. The model or framework should be molded to each culture that it is being forced into. In essence, we must all play with the hand we are dealt. We do the best we can with what we have.

To that end, the process flow involved with such team activities might look like Figure 7.1.

We can speak ideally about how RCA teams should function, but rarely are there ideal situations in the real world. We have discussed throughout this text the effects of reengineering on corporate America and how resources and capital are tight while financial expectations rise. This environment does not make a strong case for organizing teams to analyze why things go wrong.

So let's bring this subject to an end with a reality check of how RCA teams will perform under the described conditions. Let's review the analysis of chronic events and how teams will realistically deal with them. Remember that the chronic events are typically viewed as acceptable, they are part of the budget, and they generally do not hurt people or cause massive amounts of damage to equipment. However, they cost the organization the most in losses on an annual basis.

Assume that a modified failure modes and effects analysis (FMEA) has been performed, and the "significant few" candidates have been determined. These will likely be chronic vs. sporadic events. Now a team has been formed using the principles outlined in this chapter. Where do we go from here?

The first meeting of an RCA team should be to define the structure of the team and delineate the team's focus. As described in this chapter, the team should first meet to develop their charter, critical success factors (CSFs), and the anticipated start and completion dates for the analysis. This session will usually last anywhere from one to two hours. At the conclusion of this meeting, the team should set the next meeting date as soon as possible.

As discussed earlier, because the nature of the events are chronic, we have in our favor frequency of occurrence. From a data collection standpoint, this means opportunity because the event is likely to occur again. Knowing that the occurrence is likely to happen again, we can plan to collect data about the event. This brings us to the second meeting of the team, to develop a data collection strategy as described in Chapter 6. Such a brainstorming session should be the second meeting of the team. This meeting typically will take only about one to two hours and should be scheduled when convenient to the team members' schedules. The result of this meeting will be assignments for members to collect various information by a certain date. At the end of this meeting, the next meeting should be scheduled. The time frame will be dependent on when the information can be realistically collected by.

The next meeting will be the first of several involving the delineation of logic using the logic tree described in Chapter 8. These sessions are reiterative and involve the thinking out of cause-and-effect or error-change relationships. The first meeting of the logic tree development will involve about two hours of developing logic paths. It has been our experience that the team should only drive down about three to four levels on the tree per meeting. This is typically where the necessary data begins to dwindle and hypotheses require more data in order to prove or disprove them. The first tree-building session will incorporate the data collected from the team's brainstorming session on data collection. The entire meeting usually takes about four hours. We find that about two hours is spent on developing the logic tree and another two hours is spent on applying verification information to each hypothesis. At the conclusion of this meeting, a new set of assignments will emerge where verification tests and completion dates will have to be applied to prove or disprove hypotheses. At the conclusion of such logic-tree-building sessions, the next meeting date should be set based on the reality of when such tests can be completed.

Our typical logic tree spans anywhere from 10 to 14 levels of logic. This coincides with the error-change phenomenon described in Chapter 6. This means that approximately three to four logic-tree-building sessions will be required to complete the tree and arrive at the root causes. To recap, this means that the team will meet on an as-needed basis three to four times for about four hours each in order to complete the logic tree. We are trying to disprove the myth that such RCA teams are taken out of the field full time for weeks on-end. We do not want to mislead at this point; we are talking time spent with team members meeting with each other. This is minimal time relative to the time required in the field to actually collect their assigned data and perform their required tests. Proving and disproving hypotheses in the field is by far the most time-consuming task in such an analysis. But it is also the most important task if the analysis is to draw accurate conclusions.

By the end of the last logic-tree-building meeting, all the root causes have been identified and the next meeting date has been set. The next meeting will involve the assigning of team members to write recommendations or countermeasures for each identified root cause. The teams as a whole will then review these recommendations; they will then strive for consensus. At the conclusion of this meeting, the last and final meeting date will be set.

The last team meeting will involve the writing of the report and the development of the final presentation. This meeting may require at least one day because we are preparing for our day in court and we want to have our solid case ready. Typically the principal analyst will have the chore of writing the report for review by the entire team. The team will work on the development of a professional final presentation. Each team member should take a role in the final presentation to show unity in purpose for the team as a whole. The development of the final report and presentation will be discussed at length in Chapter 9.

As we can see, we have to deal with the reality of our environments. Keep in mind that the above-described process is an average for a chronic event. If someone in authority pinpoints any event as a high priority, this process tends to move much faster as support tends to be offered rather than fought for.

We will now move into the details of actually taking the pieces of the puzzle (the data collected) with the ideal team assigned and make sense of a seemingly chaotic situation.

8 Analyze the Data: Introducing the Logic Tree

No matter what methods are out in the marketplace to conduct root cause analysis (RCA), they all have one thing in common: cause and effect relationships! This is the aspect of science that makes finding root causes possible. The various RCA methods in the marketplace may vary in presentation, but the legitimate ones are merely different in the way in which they graphically represent the cause and effect relationships. Everyone will have their favorite tool, which is fine, as long as they are using it and it is producing results.

In this chapter we are going to describe our RCA method of choice called a *logic tree*. This is our means of organizing all the data collected thus far and putting it into an understandable and logical format for comprehension. This is different from the traditional logic diagram and a traditional fault tree. A logic diagram is typically a decision flow chart that will ask a question, and depending on the answer, will direct the user to a predetermined path of logic. Logic diagrams are popular in situations where the logic of a system has been laid out to aid in human decision making. For instance, an operator in a nuclear power generation facility might use such a logic diagram when an abnormal situation arises on the control panel and a quick response is required.

A fault tree is traditionally a totally probabilistic tool that uses a graphical tree concept that starts with a hypothetical event. For instance, we may be interested in how that event could occur, so we would deduce the possibilities on the lower level.

A logic tree is a combination of both of the above tools. The answer to certain questions will lead the user to the next lower level. However, the event and its surrounding modes will be factual vs. hypothetical. The basic logic tree architecture is shown in Figure 8.1.

We will begin to dissect the above architecture to gain a full understanding of each of its components in order to fully understand its power.

LOGIC TREE ARCHITECTURE

THE EVENT

This is a brief description of the undesirable outcome that is being analyzed. This is an extremely important block because it sets the stage for the remainder of the analysis. *This block must be a fact.* It cannot be an assumption. From an equipment standpoint, the event is typically the loss of function of a piece of equipment and/or

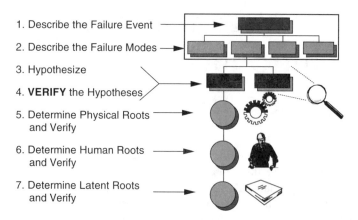

1. Describe the Failure Event

2. Describe the Failure Modes

3. Hypothesize

4. **VERIFY** the Hypotheses

5. Determine Physical Roots and Verify

6. Determine Human Roots and Verify

7. Determine Latent Roots and Verify

FIGURE 8.1 Logic tree architecture.

Recurring Pump Failure

FIGURE 8.2 Event example.

process. From a production standpoint, it is the reason that the organization cares about the undesirable outcome. Under certain conditions we will accept such an undesirable outcome, whereas under other conditions we will not.

Let us take an example of the above statement. When we are in a business environment that is not sold out (meaning we cannot sell all we can make), we are more tolerant of equipment failures that restrict capacity because we do not need the capacity anyway. However, when sales pick up and the additional capacity is needed, we cannot tolerate such stoppages and rate restrictions. In the non-sold-out state, the event may be accepted. In the sold-out state, it is not accepted. This is what we meant by the event being defined as "the reason we care." We only cared here because we could have sold the product for a profit.

Please remember our earlier discussion about the error-change phenomenon in Chapter 6. We discussed how error-change relationships are synonymous with cause-effect relationships. The event is essentially the last link in the error chain. It is the last effect and usually the reason we notice that something is wrong. See Figure 8.2 for an example of an event.

THE EVENT MODE(S)

The modes are a further description of how the event *has* occurred in the past. *Remember, the event and mode levels must be facts.* This is what separates the logic tree from a fault tree. It is a deduction from the event block and seeks to break down the bigger picture into smaller, more manageable blocks. Modes are typically easier to delineate when analyzing chronic events. Figure 8.3 illustrates a typical top box (event plus mode level) for a chronic event.

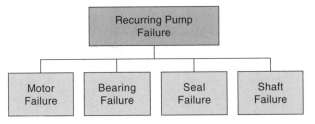

FIGURE 8.3 Top box example of chronic event.

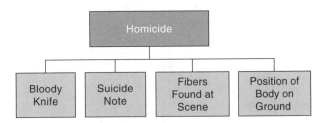

FIGURE 8.4 Top box example for sporadic event.

When dealing with sporadic events (one-time occurrences), we do not have the luxury of repetition so we must rely on the facts at the scene. This is a direct parallel to our police detective analogy in Chapter 6. When a homicide occurs, the evidence at the scene is critical and hence becomes our modes. Modes in sporadic incidents are the observations that are abnormal and need to be explained. To capitalize on our detective analogy, Figure 8.4 shows a top box example for a homicide scene.

THE TOP BOX

The top box is the aggregation of the event and the mode levels. As we have emphatically stated, *these levels must be facts!* We state this because it has been our experience that the majority of the time we deal with RCA teams, there is a propensity to act on assumptions as if they are facts. This assumption and subsequent action can lead an analysis in a completely wrong direction. The analysis must begin with facts that are verified, conventional wisdom, ignorance, and opinion should not be accepted as fact.

To illustrate the dangers of accepting opinion as fact, I will relate a scenario we encountered. We were hired by a natural gas processing firm to determine how to eliminate a phenomenon called *foaming* in an amine scrubbing unit used in their process. In order to get the point across about the top box and not have to get into a technical understanding, the illustration in Figure 8.5 would be a basic drawing of a scrubbing unit (very similar to a distillation column) with bubble-cap trays and downcomers. The purpose of this vessel is to clean and sweeten the gas for the gas producers and make it acceptable for the gas producers' customers.

The event described by the company that hired us was vehemently *foaming*. Foaming is a phenomenon by which a foam is formed within the amine scrubbing

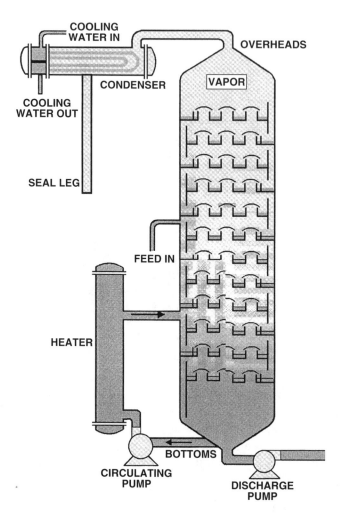

FIGURE 8.5 Amine scrubbing unit illustration.

unit which restricts the flow of gas through the bubble-cap trays. As a result, capacity is restricted and they are unable to meet customer demand due to unreliability of the process. When asked how long this had been occurring, the reply was 15 years! Why would an organization accept such an event for 15 years? The answer was candid and simple: over the past 15 years there was more capacity than demand. Therefore, rate restrictions were not costing as much money. Now the business

DISTILLATION PROCESS

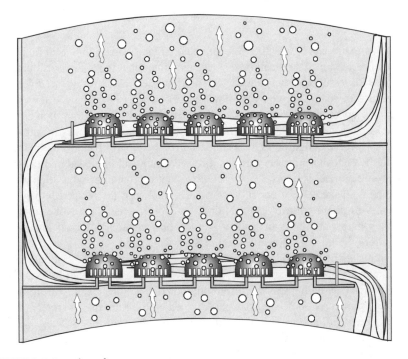

FIGURE 8.5 (continued)

environment has changed where demand has outpaced capacity and the facility cannot meet the challenge. Now it is costing them lost profit opportunities.

Given the above scenario, what is the event and what is the mode? The natural tendency of the team of experts was to label the event as *foaming*. After all, they had a vested interest in this label as all of their corrective actions to date were geared toward eliminating foaming. But what are the facts of the scenario? We were unbiased facilitators of our RCA process, therefore we could ask any question we wished. What was the real reason they cared about the perceived foaming event? They were only concerned now because they could not meet customer demand. That is the fact! This analysis would not have taken place if they had not been getting complaints and threats from their customers.

Given the event is "recurring process interruptions prevent ability to satisfy customer demand," what is the mode? What is the symptom of why the event is occurring? At this point the natural tendency of the team again was to identify

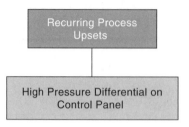

FIGURE 8.6 Foaming example of top box.

foaming as the mode. Remember, modes must be facts. Being unbiased facilitators and not experts in the technical process, we explained that the amine scrubbing unit was a closed vessel. In other words, we could not see inside the vessel to confirm the presence of foam. So we asked how they know they have foam if they cannot see it. This question seemed to stifle the team for a while as they pondered the answer. Minutes later, one of the operators replied that they know they have foaming when they receive a high pressure differential on the control panel. The instrumentation in the control room was calibrated for accuracy and indicated that the instruments were accurate. The FACT in this case was not yet foaming, but a "high pressure differential" on the control panel. This was the indicator that leaped into people's minds, having them believe that foaming did indeed exist. This is a very typical situation where our minds make leaps based on indications. Based on this new information, the top box shown in Figure 8.6 was created.

Coming off of this mode level, we would begin to hypothesize about how the preceding event could have occurred. Therefore, our question would become "How could we have a high pressure differential on the control panel resulting in a restriction of the process?" The answers supplied by the expert team members were: foaming, fouling, flooding, and plugged coalescing filters upstream. These were the only ways they could think of to cause a high pressure differential on the board. Now came the task of verifying which were true.

We simply asked the question, "How can we verify foaming?" Again, a silence overcame the crowd for about 15 seconds until one team member rose and stated that they had taken over 150 samples, and they could not get any to foam. In essence, they had disproved that foaming existed but would not believe it because it was the only logical explanation at the time. They honestly believed that foam was the culprit and acted accordingly.

To make a long story short, the operators received the indicator that a high pressure differential existed. Per their experience and education, they responded to the indicator as a foaming condition. The proper response under the assumed conditions was to shoot into the vessel a liquid called antifoam which is designed to break down foam (if it existed). The problem was that no one knew exactly how much antifoam to shoot in nor how much they were shooting in. It turned out that the operators were shooting in so much anti-foam they were flooding the trays. They were treating a condition that did not exist and creating another condition that restricted flow.

The original high pressure differential was determined to be caused by a screen problem in a coalescing filter upstream. But no one considered that condition as an

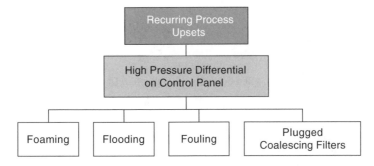

FIGURE 8.7 Foaming top box and first level example.

option at the time. The point to the whole story is that if we had accepted the team's opinion of foaming as fact, we would have pursued a path that was incorrect. This is the reason that we are vehement about making sure the top box is factual. The resulting top box and first hypotheses looked like the example shown in Figure 8.7.

THE HYPOTHESES

As we learned in school at an early age, a hypothesis is merely an educated guess. Without making it any more complex than that, hypotheses are responses to the "How can?" questions described previously. For instance, in the foaming example we concluded that the high pressure differential was the mode. This is the point at which the facts stop and we must hypothesize. At this point we do not know why there is a high pressure differential on the board, we just know that it exists. So we simply ask the question, "How could the preceding event have occurred"? The answers we seek should be as broad and all inclusive as possible. As we will find in the remainder of this chapter, this is contrary to normal problem-solving thought processes.

VERIFICATIONS OF HYPOTHESES

Hypotheses that are accepted without validation are merely assumptions. This approach, though a prevalent problem-solving strategy, is really no more than a trial and error approach. In other words, it appears to be this case, so we will spend money on this fix. When that does not work, we reiterate the process and spend money on the next likely cause. This is an exhaustive and expensive approach to problem solving.

In the PROACT® methodology, all hypotheses must be supported with hard data. The initial data for this purpose was collected in our 5-Ps effort in the categories of parts, position, people, paper, and paradigms. The 5-Ps data will ultimately be used to validate hypotheses on the logic tree. While this is a vigorous approach, the same parallel is used for the police detective preparing for court. The detective seeks a solid case, and so do we. A solid case is built on facts, not assumptions. Would we expect a detective to win a murder case with a drug dealer as the only witness? This is a weak case and not likely to be successful.

TABLE 8.1
Sample Verification Log

Hypothesis	Verification Method	Responsibility	Completion Date	Outcome	Confidence
Foam	Perform foam tests on samples	RHT	09/09/99	False	0
Flooding	Monitor volume of antifoam in scrubber	TGJ	09/10/99	True	5
Fouling	Inspect trays	RPG	09/11/99	False	0
Plugged coalescing filters	Inspect filters	RCA	09/12/99	True	5

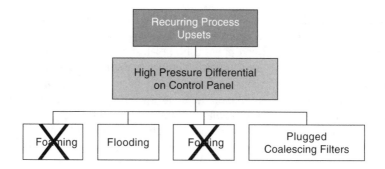

FIGURE 8.8 The fact line positioning

Hard data for validation means eyewitness accounts, statistics, certified tests, inspections, online measurement data, and the like. A hypothesis that is proven to be true with hard data becomes a fact. In keeping with our solid case analogy, we must keep in mind that organization is a key to preparing our case. To that end, we should maintain a verification log on a continuing basis to document our supporting data. Table 8.1 provides a sample of a verification log used in RCA. This document supports the tree and allows it to stand up (especially in court).

THE FACT LINE

The fact line starts below the mode level because above it are facts and below it are hypotheses. As hypotheses are proven to be true and become facts, the fact line moves down the length of the tree. For instance, see Figure 8.8 for an illustration of the fact line for the foaming example.

PHYSICAL ROOT CAUSES

The first root level causes that are encountered through the reiterative process will be the physical roots. Physical roots are the tangible roots or component level roots. In many cases, when undisciplined problem-solving methods are used, people will have a tendency to stop at this level and call them root causes. We do not subscribe to this type of thinking. In any event, all physical root causes must also endure validation to prove them as facts.

HUMAN ROOT CAUSES

Human root causes will almost always trigger a physical root cause to occur. Human root causes are decision errors that result in errors of omission or commission. This means that either we decided not to do something we should have done, or we did something we were not supposed to do. Examples of errors of omission might be that we were so inundated with reactive work, we purposely put needed inspection work on the back burner. An error of commission might be that we aligned a piece of equipment improperly because we did not know how to do it correctly.

While the questioning process thus far has been consistent with asking "How can?" at the human root level we want to switch the questioning to "Why?" When dealing in the physical and process areas, it is not appropriate to ask equipment why something occurs. Only at the human root level do we encounter a person. When we get to this level, we are not interested in "whodunnit" but rather why they made the decision that they did. Understanding the rationale behind decisions that result in error is the key to conducting true RCA. Anyone who stops an RCA at the human level and disciplines an identified person or group is participating in a witch hunt. Witch hunts were discussed in the preserving failure data section and proven to be non-value-added as the true roots cannot be attained in this manner. This is because if we search for a scapegoat, no one else will participate in the analysis for fear of repercussions. When we cannot find out why people make the decisions that they do, we cannot solve the issue at hand.

LATENT ROOT CAUSES

Latent root causes are the organizational systems that people use to make decisions. When such systems are flawed, they result in decision errors. The term *latent** is defined as "…whose adverse consequences may lie dormant within the system for a long time, only becoming evident when they combine with other factors to breach the system's defenses."

When we use the term *organizational* or *management systems,* we are referring to the rules and laws that govern a facility. Examples of organizational systems might include policies, operating procedures, maintenance procedures, purchasing practices, stores and inventory practices, etc. These systems are all put in place to help people make better decisions. When a system is inadequate or obsolete, people

---

* Reason, James. *Human Error.* Victoria: Cambridge University Press, 1990–1992. p. 173.

FIGURE 8.9 The expert's logic tree.

end up making decision errors based on flawed information. These are the true root causes of undesirable events.

BUILDING THE LOGIC TREE

We have now defined the most relevant terms associated with the construction of a logic tree. Now let's explore the physical building of the tree and the thought processes that go on in the human mind.

Experts who participate on such teams are generally well educated individuals, well respected within the organization as problem solvers and people who pay meticulous attention to detail. With all this said and done, using the logic tree format, an expert's thought process may look like Figure 8.9.

This poses a potential hurdle to a team, because for the most part, the analysis portion is bypassed and we go straight from problem definition to cause. It is the principal analyst's responsibility to funnel the expertise of the team in a constructive manner without alienating the team members. Such an RCA team will have a tendency to go to the micro view and not the macro view. However, in order to understand exactly what is happening, we must step back and look at the big picture. In order to do this, we must derive exactly where our thought process originated from and search for assumptions in the logic.

A logic tree is merely a graphic expression of what a thought would look like if it were on paper. It is actually looking at how we think. Let's take a simple example of a pump of some type that is failing. We find that 80% of the time this pump is failing due to a bearing failure. This shall serve as the mode that we pursue for demonstration purposes (see Figure 8.10).

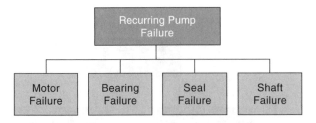

FIGURE 8.10 Recurring pump failure example.

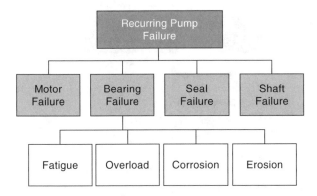

FIGURE 8.11 Broad and all-inclusive thinking.

BROAD AND ALL INCLUSIVENESS

If we have a team of operations, maintenance, and technical members and ask them how a bearing could fail their answers would likely get into the nuts and bolts of such details as it was installed improperly, it was a design error, defective materials, too much or too little lubricant, misalignment, and the like. While these are all very valid, they jump to too much detail too fast. We want to use deductive logic in short leaps.

In order to be "broad and all inclusive" at each level, we want to possess all the possible hypotheses in the fewest amount of blocks. To do this, we must think as if we are the part being analyzed. For instance, in the above example with the bearing, if we thought of ourselves as being the bearing, we would think about "How exactly did we fail?" From a physical failure standpoint, the bearing would have to erode, corrode, overload, or fatigue. These are the only ways the bearing can fail. All of the hypotheses developed by the experts earlier would cause one or more of these states to occur. See Figure 8.11.

From this point we would have a metallurgical review of the bearing conducted. If the results were to come back and state that the bearing failed due to fatigue, then there are only certain conditions that can cause a fatigue failure to occur. The data leads us in the correct direction, not the team leader. This process is entirely data driven.

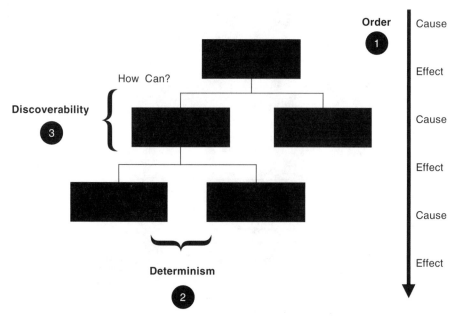

FIGURE 8.12 The three keys.

If we are broad and all-inclusive at each level of the logic tree and we verify each hypothesis with hard data, then the fact line drops until we have uncovered all the root causes. This is very similar to the recent quality initiatives; we are ensuring quality of the process so that by the time the root causes are determined, they are correct.

THE ERROR-CHANGE PHENOMENON APPLIED TO THE LOGIC TREE

Now let's explore how the error-change concept (cause-effect relationship) parallels the logic tree. As we explore the path of the logic tree, there are three key signs of hope in favor of our finding the true root causes (see Figure 8.12). These keys are:

1. Order
2. Determinism
3. Discoverability

ORDER

If we truly believe that the error-change phenomenon exists, then we have the hope that following cause and effect relationships backwards will lead us to the culprits, the root causes. We often ask our classes if they believe that there is order in everything, including nature. There is generally a silent pause until they think about it and they cite facts such as tides coming in and going out at predetermined times, the sun rising and setting at predetermined times, and the seasons that various geographic regions experience on a cyclical basis. These are all indications that such order, or patterns, exist.

DETERMINISM

This means that everything is determinable or predictable within a range. If we know that a bearing has failed, the reasons (hypotheses) of how the bearing can fail are determinable. We discussed this earlier with the options being corrosion, erosion, fatigue, and overload. This is determinable within a range of possibilities.

People are the same way to a degree. People are determinable within a broader range than equipment because of the variability of the human race. If we subject humans to a specific stimuli, they will react within a certain range of behaviors. If we alienate employees publicly, then chances are they will withdraw their ability to add value to their work. They in essence become human robots because we treated them that way.

Determinism is important because when constructing the logic tree it becomes essential, from level to level, to develop hypotheses based on determinism.

DISCOVERABILITY

This is the simple concept that when you answer a question, it merely begets another question. We like to use the analogy of children in the ranges of three to five years old. They make beautiful principal analysts because of their inquisitiveness and openness to new information. We have all experienced our children at this age when they say "Daddy, why does this happen?" We can generally answer the series of "why" questions about five times before we do not know the actual answer. This is discoverability — questions only lead to more questions. On the logic tree, discoverability is expressed from level to level when we ask "How can something occur?" The answer only leads to another "How can?" question.

All of these keys provide the analyst with the hope that there is a light at the end of the tunnel and it is not a train. We are basically searching for pattern in a sea of chaos and the three keys help us find pattern in the chaos.

Imagine if we were the investigators at the bombing site of the Twin Towers building in New York years ago. Could we even visualize finding the answer from looking at the rubble generated from blast; the chaos? Yet, the investigators knew that there was a pattern in the chaos somewhere and they were going to find it. Apparently, within two weeks of the blast, the investigators knew the type of vehicle, the rental truck agency, and the makeup of the bomb. This is true faith in finding pattern in chaos. These people believe in the logic of failure.

AN ACADEMIC EXAMPLE

Let's take all of the described pieces of the logic tree architecture and now put them into perspective in an academic example we can all relate to.

We have all experienced problems, at some time or another, with our local area networks (LANs) at our offices. This is a very universal issue we see in our travels.

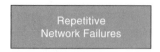

FIGURE 8.13 The LAN event block.

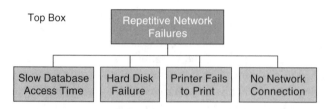

FIGURE 8.14 LAN example top box.

If we were to look at this issue from an RCA perspective, the question would be: "What is the event in this case?" The end of the error chain is that the LAN is not functioning as designed. Therefore, we may want to paraphrase and say recurring LAN failures is our event because it is the reason we care, the last effect of the cause and effect chain. See Figure 8.13.

Now let's move to the second level and describe the modes that are the symptoms of how we know that the LAN has not been performing as designed. This information, in this situation, may come from users at their workstations in the form of complaints to the information systems (IS) department. Some examples of modes at this level might be slow database access time, hard disk failure, printer fails to print, and no network connection. These are all facts that have been observed by the users in the past.

Now which mode would we want to approach first? Had we performed a modified failure modes and effects analysis (FMEA), we would already have the event and mode that has been the most costly to the organization. In this example we are going to pursue the mode with the greatest frequency of occurrence and that would be that the printer fails to print. In this particular office, the majority of the complaints have been that the printer does not print when they send a job. These complaints absorb about 80% of the IT technicians' time. For this reason, we will pursue this leg. The top box may look like Figure 8.14.

At this point we begin our hypothetical questioning into how the printer could fail to print. The natural thought process would be to respond with such answers as no toner, no power, wrong configuration, operator error, no paper, etc. All of these hypotheses are valid, but do they meet the criteria of broad and all inclusive? This is the most difficult portion of constructing a logic tree, thinking broadly! Is it possible that all of the possibilities would have to be embedded somewhere in the printer, the computer, or the cable? We will use the possibility of the operator in a verification, but these three hypotheses provide broadness and all-inclusiveness. The next level of this logic tree would look something like Figure 8.15.

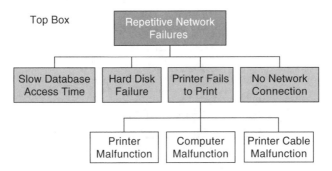

FIGURE 8.15 The first hypothetical leg.

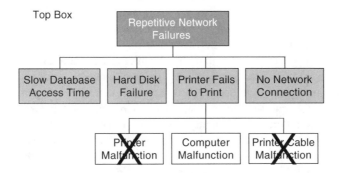

FIGURE 8.16 Updated logic tree.

Now comes the task of proving which hypotheses are true, and which are not. It is at this point that the verification log begins to be developed and that we use information collected in our 5-Ps as validation data. Let's take the first hypothesis of the printer and determine a test that can prove or disprove it. We can take our laptop computer and connect it to this printer with the same cable and the same operator to test its functionality. In this case, the printer functions as designed. Based on this test, we can cross out the hypotheses of the printer and the cable (and also the theory of the operator). However, we cannot select the computer by process of elimination. The computer must also have a test to validate it. In this case we can connect another known working printer to the same computer to test its functionality with the same operator. We conclude from this test that the new printer also does not perform with the same computer. Based on these tests, the logic tree would look like Figure 8.16.

At this point our fact line has moved down from the mode level to the first hypothesis level. Because the hypothesis of the computer has been verified as true, it is now a fact and the fact line drops.

We continue our questioning by pursuing the next level and asking "How could a computer malfunction cause the printer not to print?" This is the discovery portion

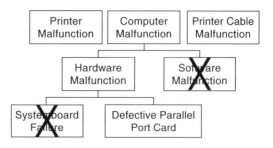

FIGURE 8.17 Hypothesis validation.

of the logic tree where one question only begets another question. Again, a hundred reasons could be thought of as to how a computer could malfunction, but we need to think broadly. The broadest blocks we can think of are hardware malfunction and software malfunction. Now tests must be developed to prove or disprove these hypotheses. Running diagnostic software determines that the system is not recognizing a parallel port card. Other than the identified hardware malfunction, there are no indications of any software malfunctions. This allows us to cross out software malfunction and continue to pursue hardware malfunction. See Figure 8.17.

The reiterative questioning continues with "How can we have a hardware malfunction that would create a computer malfunction that would not allow the printer to print?" We notice in this questioning that we are always reading the logic path back a few levels to maintain the story or error path. This helps the team follow the logic tree and put the question into proper perspective. Our broad and all-inclusive answer here could be either a system board failure or a parallel port card malfunction. The previous test of running the diagnostic software confirms an issue with the parallel port card. The system board, or the mother board as the computer jockeys call it, has displayed no signs of malfunction with the context of the entire system. If a problem was apparent with the system board, there would be more apparent issues other than just a printer failing to print. The absence of these issues is the validation that the system board does not appear to be a contributor to this event.

At this point we remove the parallel port card, clean the contact areas, and reseat it back into the appropriate slots making sure the contact is being made and that improper installation concerns are not an issue. The printer still fails to print even when correct installation of the parallel port card has been confirmed. Next we replace the parallel port card with one known to be working and properly install it into the computer. This time the printer works as desired. Many feel the analysis is complete at this level because the event will not occur immediately. However, this is a point that we consider it a physical root when the event temporarily goes away. For this reason we would circle the block identified as a defective parallel port card, indicating it as a root cause. See Figure 8.18.

Having identified the physical root in this case means that we have more work to do in order to determine the latent roots. Our questioning continues with "How can we have a defective parallel port card which is causing a hardware malfunction which is causing a computer malfunction which is causing the printer not to print?"

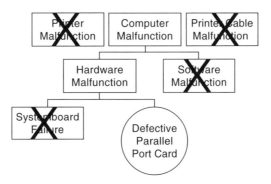

FIGURE 8.18 Identification of physical root cause.

Either we installed it improperly or we purchased it in a defective state, or it became defective while in our use.

We have already determined that the installation practice was proper. We can eliminate that the card became defective in our possession because interviews reveal that this was a new printer being added to the network and that it never worked from the beginning. People on the network therefore chose to divert to another network printer. So this was not a case where the printer worked at one time and then did not work. This serves as the proof that the parallel port card did not become defective in our possession but rather that we received it that way from the manufacturer.

Now we must review our purchasing practices and determine if we have purchasing procedural flaws that are allowing defective parts to enter the organization. From our 5-Ps information, we determine that there is no list of qualified vendors, and we have inadequate component specifications. We find that the primary concern of purchasing is to buy based on low cost because that is where the incentives are placed for the purchasing agents in this firm. We have now confirmed that we purchased a poor quality card and because this task involves a conscious human decision that results in an action to be taken, this is deemed our human root cause. We circle this block now designating it as such. Remember at this human root level, our questioning switches to Why? because we have reached a human being that can respond to the question. Our question at this point would read, "Why did we choose to purchase this particular parallel port card from another vendor?" Our answers are "no list of qualified vendors" and "no component specifications" and "low cost mentality." These are the reasons behind the decision and therefore are the latent roots. See Figure 8.19.

This example, although academic, could relate to situations in our own environments where disruptions are caused in our processes due to the infiltration of defective parts into the organization. Without a structured RCA approach, we would use trial and error approaches until something worked. This is very expensive.

What if we stopped at the physical root of "defective parallel port card" and just replaced the card? Would the event likely recur? Sure it would if the same purchasing habits continue. It may not happen in the same location because not all cards would be defective, but it would likely happen somewhere else in the organization, forcing another need to problem solve.

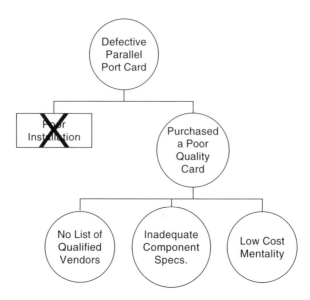

FIGURE 8.19 Identification of human and latent roots.

What if we stopped at the human root of "purchased a defective parallel port card" and disciplined the purchasing agent who made the decision? Would that prevent recurrence? Likely not because the decision-making system that the agent used is likely being used by other purchasing agents in the organization. It might prevent that agent from making such a decision in the future, but it would not stop other such decisions from being made in the future by other agents.

The only way to prevent recurrence of this event in the entire organization is to correct the decision-making systems we refer to as the latent roots or organizational system deficiencies. When such deficiencies are uncovered then we are truly performing "root" cause analysis.

The completed verification log for the above example might look like Table 8.2.

VERIFICATION TECHNIQUES

While we used simple verification techniques in the above example, there are thousands of ways in which to validate hypotheses. They are all, obviously, dependent on the nature of the hypothesis. The following is a list of some common verification techniques used in industrial settings:

- Human observation
- Fractology
- High speed photography
- Video cameras
- Laser alignment
- Vibration monitoring and analysis
- Ultrasonics

TABLE 8.2
Completed Sample of Verification Log

Hypothesis	Verification Method	Responsibility	Completion Date	Outcome	Confidence
Printer malfunction	We used a stand-alone laptop to test the printer.	RMB	09/01/99	Printer worked fine on alternate computer	0
Computer malfunction	Use a known working printer and test it on the computer	TDF	09/01/99	Still could not print after the test	5
Printer cable malfunction	We used a stand-alone laptop to test the cable.	RMB	09/01/99	Cable worked fine on alternate computer	0
Hardware malfunction	Run diagnostic software to check hardware	TDF	09/01/99	Determined a possible problem with parallel port card	5
Software malfunction	Check drivers and configuration	TDF	09/02/99	Configuration and drivers were correct	0
Systemboard failure	Call in a technician to test the system board for faults	TDF	09/03/99	Not indication of systemboard failure	0
Defective parallel port card	Replace the card with a card known to work	RMB	09/04/99	The document printed fine using the alternate card	5
Poor installation	Check installation notes as well as talking with technician who installed the card	RMB	09/03/99	Installation looked adequate	0
Purchased a poor quality card	Talk with the purchasing department and storeroom personnel	RMB	09/03/99	Determined that this was a new installation and discovered the card never worked properly	5
No list of qualified vendors	Determine current vendor requirements	JCF	09/04/99	Records determined that we have no list of qualified vendors	5
Low cost mentality	Look for a history of low bidder mentality	FRD	09/03/99	This has been the prevalent purchasing practice in the purchasing department	5
Inadequate component specifications	Check the component specifications for this and other related items	FGH	09/04/99	We did not have solid component specifications	5

- Eddy current testing
- Infrared thermography
- Ferrography
- Scanning electron microscopy
- Metallurgical analysis
- Chemical analysis
- Statistical analysis (correlation, regression, Weibull Analysis, etc.)
- Operating deflection shapes
- Finite element analysis modeling
- Motor signature analysis
- Modal analysis
- Experimental stress analysis
- Rotor dynamics analysis
- Capacity and Availability Assessment Program (CAAP®)*
- Work sampling
- Task analysis

These are just a few to give us a feel for the breadth of verification techniques that are available. There are literally thousands more. Each of these topics could be a text in itself. Many texts are currently available to provide us more in-depth knowledge on each of these techniques. However, the focus of this text is on the root cause analysis method. A good principal analyst does not necessarily have to be an expert in any or all of these techniques; rather they should be resourceful enough to know when to use which technique and how to obtain the resource to complete the test. A principal analyst should have a repository of resources to tap into when the situation permits.

CONFIDENCE FACTORS

It has been our experience that the more timely, pertinent data that is collected with regards to a specific event, the quicker the analysis is completed and the more accurate the results are. Conversely, the less data we have, the longer the analysis takes and the greater the risk of the wrong cause(s) being identified.

We use a confidence factor rating for each hypothesis to evaluate how confident we are with the validity of the test and the accuracy of the conclusion. The scale is basic and runs from 0 to 5. A 0 means without a doubt, with 100% certainty, that with the data collected the hypothesis is *not* true. On the flip side, a 5 means that with the data collected and the tests performed, there is 100% certainty that the hypothesis is true. Between the 0 and the 5 are the shades of gray where the data used was not absolutely conclusive. This is not uncommon in situations where an RCA is commissioned weeks after the event occurred and little or no data from the scene was collected. Also, we have seen in catastrophic explosions where uncertainty resides in the physical environment prior to the explosion. What formed the combustible environment? These are just a few circumstances in which absolute certainty

---

* CAAP is a registered trademark of Applied Reliability Incorporated (ARI), Baton Rouge, LA, 1998.

cannot be attained. The confidence factor rating communicates this level of certainty and can guide corrective action decisions.

We use the rule of thumb that with a confidence rating of 3 or higher, we treat the hypothesis as if it did happen and pursue the logic leg. Any confidence rating of less than 3 we treat as a low probability of occurrence and feel that it should not be pursued at this time. However, the only hypotheses that are crossed out are the ones that have a confidence rating of 0. A 1 cannot be crossed out because it still has a probability of occurring even if the probability was low.

THE TROUBLESHOOTING FLOW DIAGRAM

Once the logic tree is completed, it should serve as a troubleshooting flow diagram for the organization. Chances are that the causes identified in this RCA will affect the rest of the organization. Therefore, some recommendations will be implemented sitewide. To optimize the use of a world class RCA effort, the goal should be the development of a dynamic troubleshooting flow diagram repository. These will end up being logic diagrams that capture the expertise on the organization's best problem solvers on paper.

Such logic diagrams can be stored on the company's intranet and available to all facilities that have similar operations and can learn from the work done at one site. These logic diagrams are complete with test procedures for each hypothesis. It is dynamic because where this RCA team may not have followed one particular hypothesis (because it was not true in their case), it may be true in another case and the new RCA team can pick up from that point and explore the new logic path.

The goal of the organization should be to capture the intellectual capital of the workforce and make it available for all to learn from. This is optimizing the intellectual capital of the organization through RCA.

THE LOGIC TREE APPLIED TO CRIMINOLOGY

Since we have been using the detective analogy throughout this process, we thought it would be appropriate to do so with the logic tree as well.

To quickly refresh our memory to this point, we strived to preserve the event data using the 5-Ps. We assembled the appropriate team in the ordering the analysis team section and then we used the logic tree process to graphically depict our cause and effect relationships leading to the event being analyzed. How does this correspond to the role of the detective? Let's reflect back on our basic structure of the logic tree we used earlier. See Figure 8.20.

This insight became apparent one night when watching CSI (Crime Scene Investigation) on television. For those who have not seen this series, it is about how advanced sciences are used to solve crimes in the U.S. It is a very riveting and enlightening series that provides a "hats off" to those forensic scientists behind the detective's and the prosecutor's case.

Now let's explore the similarities between their work and ours. The top box is composed of the event and the modes. These have simply been described as *factual*

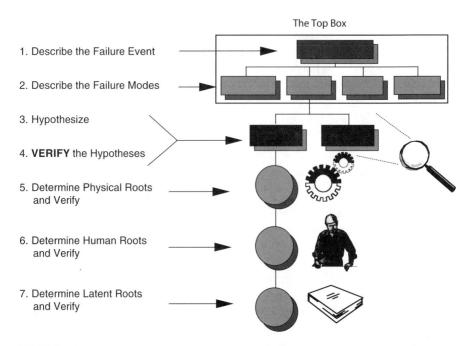

1. Describe the Failure Event

2. Describe the Failure Modes

3. Hypothesize

4. **VERIFY** the Hypotheses

5. Determine Physical Roots
   and Verify

6. Determine Human Roots
   and Verify

7. Determine Latent Roots
   and Verify

FIGURE 8.20 Logic tree architecture.

information that properly defines the event. A fact means that it is proven and not left to supposition. What do the detectives have to base their case on? Their facts generally start with the evidence at the scene of the crime. As discussed earlier, the line around the top box could be synonymous with the yellow police tape around the crime scene. The police tape could be viewed as sealing off the area and preserving the facts. The facts ultimately serve as evidence in building the case. From the facts, leads are generated (hence the name lead).

When leads are generated in these situation, it is typically because the investigators ask the question, "How could this type of trauma to the head have occurred?" This *leads* them to their hypotheses. Once their hypotheses have been developed, they must use forensic science (as in CSI) to prove or disprove them with the science that is available. This is akin to our using a metallurgical lab to use electron microscopy to view a crystalline structure of a fractured surface. Science in both cases is used to determine the *how* of the crime or the event.

When exploring the human and latent root causes, we venture off into areas that are not typically the expertise of the forensic scientists. This is because we are now looking for the "*why* did this person make the decision to commit a crime." In police terms, they are now seeking *motive*. The same goes for our facilities. The metallurgist would not be the one to determine the intent of the person who made the wrong decision; the metallurgist would state the facts about the fractured surface. The responsibility for determining the *motive* falls to the investigator and the prosecutor.

The moral of this story is that the top box is equivalent to the crime scene. The development of hypotheses and their verification are the equivalent of the *how* in the crime lab. The human and latent root causes are equivalent to establishing *motive* or the *why* someone made the decision that they did.

We just wanted to show that this logical process is not only applicable in certain situations, but in any situation. The processes of logical deduction and reasoning are elements of the human being and our ability to resolve problems. The nature of the problem at hand should be irrelevant.

9 Communicating Findings and Recommendations

THE RECOMMENDATION ACCEPTANCE CRITERIA

Let's assume at this point that the complete RCA process has been followed to the letter. We have conducted our modified FMEA and determined our "significant few." We have chosen a specific significant event and proceeded through the PROACT® process. An identified RCA team has undertaken an organized data collection effort. The team's charter and critical success factors (CSF) have been determined and a principal analyst (PA) has been named. A logic tree has been developed where all hypotheses have been either proven or disproved with hard data. Physical, human, and latent roots have been identified. Are we done?

Not quite! Success can be defined in many ways, but an RCA should not be deemed successful unless something has improved as a result of implementing recommendations from the RCA. Merely conducting an excellent RCA does not produce results. As many of you can attest, getting something done about your findings can be the most difficult part of the analysis. Oftentimes recommendations will fall on deaf ears and then the entire effort was a waste of your time and the company's money.

If we know that such hurdles will be evident, then we can also proactively plan for their occurrence. To that end, we suggest the development of recommendation acceptance criteria. We have all faced situations where we spend hours and sometimes weeks and months developing recommendations as a result of various projects only to have the recommendations turned down flat. Sometimes explanations are given and sometimes they are not. Regardless, it is a frustrating experience and it does not encourage creativity in making recommendations. We usually will tend to become more conservative in our recommendations to merely get by.

Recommendation acceptance criteria are what we call the rules of the game. Managers are typically the fiscally responsible people. They handle the company's money and, in doing so, make economic decisions as to how the money is spent. Whether these rules are written or unwritten, they define whether or not our recommendations will fly with management. We suggest asking the approving managers before we even begin to write our recommendations for the rules of the game. This is a reasonable request seeking only not to waste company time and money on non-value-added work.

A sample listing of recommendation acceptance criteria might look like this. The recommendation must:

1. *Eliminate or reduce the impact of the cause* — The goal of an RCA may not always be to eliminate a cause. For instance, if we find our scheduled shutdowns to be excessive, it would not be feasible to expect that they can be eliminated. Our goal may be to reduce the shutdown lengths (mean time to restore/MTTR) and increase the time between their occurrences (mean time between failure/MTBF).

2. *Provide a ___% return on investment (ROI)* — Most every company we have ever dealt with has a predetermined ROI. Ten years ago such ROIs were frequently around 15 to 20%. Recently, we have seen these numbers and expectations range from 50 to 100%. This indicates a risk-aversive culture where we only deal with certainty.

3. *Not conflict with capital projects already scheduled* — Sometimes we develop lengthy recommendations only to find that some plans are on the books, unbeknownst to us, that call for the mothballing of a unit, area, or activity. If we are informed of such secret plans, then we will not spin our wheels developing recommendations that do not have a chance.

4. *List all the resources and cost justifications* — Managements generally like to know that we have thought a great deal about how to execute the recommendations. Therefore, cost/benefit analyses, manpower resources required, materials necessary, safety and quality considerations, etc. should all be laid out.

5. *Have a synergistic effect on the entire system/process* — Sometimes in our working environments we have "kingdoms" that develop internally and we end up in a situation where we stifle communication and compete against each other. This scenario is common and counterproductive. Management should expect recommendations that are synergistic for the entire organization. Recommendations should not be accepted if they make one area look good at the sacrifice of other areas up- and downstream.

While this is a sample listing, the idea is that we do not want to waste our time and energy developing recommendations that do not have a chance of flying in the eyes of the decision maker. Efforts should be made to seek out such information and then frame the team's recommendations around the criteria.

DEVELOPING THE RECOMMENDATIONS

Every corporation will have its own standards in how it wants recommendations to be written. It will be the RCA team's goal to abide by these internal standards while accomplishing the objectives of the RCA team charter.

The core team members, at a predetermined team meeting time and location, should discuss recommendations. The entire meeting should be set aside to concentrate on recommendations alone. At this meeting the team should consider the recommendation acceptance criteria (if any were obtained) and any extenuating circumstances. Remember our analogy of the detective throughout this text, always trying to build a solid case. This report and its recommendations represent our day in court. In order to win the case, our recommendations must be solid and well thought out. But foremost, they must be accepted, implemented, and effective in order to be successful.

At this team meeting, the objective should be to gain team consensus on recommendations brought to the table. Team consensus is not team agreement. Team agreement means that everyone gets what he or she wants. Team consensus means that everyone can live with the recommendations. Everyone did not get all of what they wanted, but they can live with it. Team agreement is rare.

The recommendations should be clear, concise, and understandable. Always have the objective in mind of eliminating or greatly reducing the impact of the cause when writing the recommendations. Every effort should be taken to focus on the RCA. Sometimes we have a tendency to have pet projects that we attach to an RCA recommendation because it might have a better chance of being accepted. We liken this to riders on bills to be reviewed in Congress. They tend to bog down a good bill and threaten its passage in the long run. Management's first sight of "smoke" will affect the credibility of the entire RCA. When writing recommendations, stick to the issues at hand and focus on eliminating the risk of recurrence.

Also, when writing recommendations, always have a back-up alternative to every recommendation that is questionable. Sometimes the recommendations we develop might be perceived as "on the edge" of the criteria given by management. If this is the case, efforts should be taken to have a back-up recommendation, a recommendation that clearly fits within the defined criteria. One thing that we never want to happen is that an issue in which the presenters have some control stalls the management presentation. Absence of an acceptable recommendation is such an obstacle and every effort should be taken to gain closure of the RCA recommendations at this meeting.

DEVELOPING THE REPORT

The report represents the documentation of the solid case for court or, in our circumstances, the final management meeting. This should serve as a living document in that its greatest benefit will be that others learn from it so as to avoid recurrence of similar events at other sites within the company or organization. To this end, the professionalism of the report should suit the nature of the event being analyzed. We like to use the adage, If the event costs the corporation $5, then perform a $5 RCA. If it costs the organization $1,000,000, then perform a $1,000,000 type of RCA.

We should keep in mind that if RCAs are not prevalent in an organization, the first RCA report usually sets the standard. We should be cognizant of this and take it into consideration when developing our reports. Let's assume at this point that we have analyzed a "significant few" type of event and it is costly to the organization, so our report will reflect that level or degree of importance.

The following table of contents shall be our guide for the report:

1. The Executive Summary
   a. The Event Summary
   b. The Event Mechanism
   c. The PROACT® Description
   d. The Root Cause Action Matrix
2. The Technical Section
   a. The Identified Root Cause
   b. The Type of Root Cause
   c. The Responsibility of Executing Recommendation
   d. The Estimated Completion Date
   e. The Detailed Plan to Execute Recommendation
3. Appendices
   a. Recognition of All Participants
   b. The 5-P's Data Collection Strategies
   c. The Team Charter
   d. The Team Critical Success Factors
   e. The Logic Tree
   f. The Verification Logs
   g. The Recommendation Acceptance Criteria (If applicable)

Now let's review the significance and contribution of each element to the entire report and the overall RCA objectives.

1. **The Executive Summary** is just that, a summary. It has been our experience that the typical decision makers at the upper levels of management are not nearly as concerned with the details of the RCA as they are with the results and credibility of the RCA. This section should serve as a synopsis of the entire RCA, a quick overview. This section is meant for managers to review the event analyzed, the reason it occurred, what the team recommends to make sure it never happens again, and how much will it cost.
   a. **The Event Summary** is a description of what was observed from the point in time that the event occurred until the point in time that the event was isolated or contained. This can generally be thought of as a time line description.
   b. **The Event Mechanism** is a description of the findings of the RCA. It is a summary of the errors that led up to the point in time of the event occurrence. This is meant to give management a quick understanding of the chain of errors that were found to have caused the event in question.
   c. **The PROACT® Description** is a basic description of the PROACT® process for management. Sometimes management may not be aware of a formalized RCA process being used in the field. A basic description of such a disciplined and formal process generally adds credibility to the analysis and assures management that it was not a "buckshot" analysis.

TABLE 9.1
Sample Root Cause Action Matrix

Cause	Type	Recommendation	Responsibility	Implementation Date	Completion Date
Outdated start-up procedure	Latent	Assemble team of seasoned operators to develop the initial draft of a current start-up procedure that is appropriate for the current operation	RJL	10/20/99	11/15/99

d. **The Root Cause Action Matrix** is a table outlining the results of the entire analysis. This table is a summation of identified causes, overview of proposed recommendations, person responsible for executing recommendation, and estimated completion date. A sample Root Cause Action Matrix is shown in Table 9.1.

2. **The Technical Section** contains the details of all recommendations. This is where the technical staff may want to review the "nuts and bolts" of the analysis recommendations.

a. **The Identified Root Cause(s)** will be delineated in this section as separate line items. All causes identified in the RCA that require countermeasures will be listed here.

b. **The Type of Root Cause(s)** will be listed here to indicate their nature as being physical, human, or latent root causes. It is important to note that only in cases involving intent with malice should any indications be made as to identifying any individual or group. Even in such rare cases, it may not be prudent to specifically identify a person or group in the report because of liability concerns. Normally no recommendations are required or necessary where a human root is identified. This is because if we address the latent root or the decision-making basis that led to the occurrence of the event, then we should subsequently change the behavior of the individual. For instance, if we have identified a human root as misalignment of a shaft (no name necessary), then the actions to correct that situation might be to provide the individual the training and tools to align properly in the future. This countermeasure will address the concerns of the human root without making a specific human root recommendation and giving the potential perception of a witch hunt.

c. **The Responsibility of Executing the Recommendation** will also be listed so as to identify an individual or group that shall be accountable for the successful implementation of the recommendation.

d. **The Estimated Completion Date** shall be listed to provide an estimated time line for when each countermeasure will be completed, thus setting the anticipated time line of returns on investment.

   e. **The Detailed Plan to Execute Recommendation** section is generally
      viewed as an expansion of the root cause action matrix described above.
      Here is where all the economic justifications are located, the plans to
      resource the project (if required), the funding allocations, etc.
3. Appendices
   a. **Recognition of All Participants** is extremely important if our intent
      is to have team members participate on RCA teams in the future. It is
      suggested to note in this section every person who inputs any infor-
      mation into the analysis. All people tend to crave recognition for their
      successes.
   b. **The 5-Ps Data Collection Strategies** should be placed as an addendum
      or appendix item to show the structured efforts to gain access to the
      necessary data to make the RCA successful.
   c. **The Team Charter** should also be placed in the report to show that
      the team displayed structure and focus with regards to their efforts.
   d. **The Team Critical Success Factors** show that the team had guiding
      principles and defined the parameters of success.
   e. **The Logic Tree** is a necessary component of the report for obvious
      reasons. The logic tree will serve as a dynamic expert system (or
      troubleshooting flow diagram) for future analysts. This type of infor-
      mation will optimize the effectiveness of any corporate RCA effort by
      conveying such valuable information to other sites with similar events.
   f. **The Verification Logs** are the spine of the logic tree and a vital part
      of the report. This section will house all of the supporting documen-
      tation for hypothesis validation.
   g. **The Recommendation Acceptance Criteria** (if applicable), should
      be listed to show that the recommendations were developed around
      stated criteria. This will be helpful in explaining why certain counter-
      measures were chosen over others.

The report will serve as a living document. If a corporation wishes to optimize
the value of its intellectual capital using RCA, then the issuing of a formal professional report to other relevant parties is absolutely necessary. Serious consideration
should be given to RCA report distributions. Analysts should review their findings
and recommendations and evaluate who else in their organization may have similar
operations and therefore similar problems. These identified individuals or groups
should be put on a distribution list for the report so that they are aware that this
particular event has been successfully analyzed and recommendations have been
identified to eliminate the risk of recurrence. This is optimizing the use of the
information derived from the RCA.

In an information era, instant access to such documents is a must. Most corporations these days have their own intranets. This provides an opportunity for the
corporation to store these newly developed "dynamic expert systems" in an electronic
format allowing instant access. Corporations should explore the feasibility of adding
such information to their intranets and allowing all sites to access the information.

Whether the information is in paper or electronic format, the ability to produce RCA documentation quickly could help some organizations from a legal standpoint. Whether the investigator is a government regulatory agency, corporate lawyers, or insurance representatives, demonstrating that a disciplined RCA method was used to identify root causes can prevent some legal actions against the corporations as well as prevent fines from being imposed due to noncompliance with regulations. Most regulatory agencies that require a form of RCA to be performed by the organizations do not delineate the RCA method to be used, but rather ensure that one can be demonstrated upon audit.

THE FINAL PRESENTATION

This is the principal analyst's final day in court. It is what the entire body of RCA work is all about. Throughout the entire analysis, the team should be focused on this meeting. We have used the analogy of the detective throughout this text. In Chapter 6 we explained that a detective goes to great lengths to collect, analyze, and document data so the lawyers can present a solid case in court in order to obtain a conviction.

Our situation is not much different. Our court is a final management review group who will decide if our case is solid enough to merit the requested monies for implementing recommendations.

Realizing the importance of this meeting, we should prepare accordingly. Preparation involves the following steps:

1. Have the professionally prepared reports ready and accessible.
2. Strategize for the meeting by knowing your audience.
3. Have an agenda for the meeting.
4. Develop a clear, concise, professional presentation.
5. Coordinate the media to use in the presentation.
6. Conduct "dry runs" of the final presentation.
7. Quantify the effectiveness of the meeting.
8. Prioritize recommendations based on impact and effort.
9. Determine the "next step" strategy.

We will address each of these individually and in some depth to maximize the effectiveness of the presentation and ensure that we get what we want.

HAVE THE PROFESSIONALLY PREPARED REPORTS READY AND ACCESSIBLE

At this stage the reports should be ready, in full color, and bound. Have a report for each member of the review team as well as a copy for each team member. Part of the report includes the logic tree development. The logic tree is the focal point of the entire RCA effort and should be graphically represented as so. The logic tree should be printed on blueprint size paper, in full color, and laminated if possible. The logic tree should be proudly displayed on the wall in full view of the review committee. Keep in mind that this logic tree will likely serve as a source of pride for the management to show other divisions, departments, and corporations how progressive their area is in conducting RCA. It will truly serve as a trophy for the organization.

STRATEGIZE FOR THE MEETING BY KNOWING YOUR AUDIENCE

This is an integral step in determining the success of the RCA effort. Many people believe that they can develop a top-notch presentation that will suit all audiences. This has not been our experience. All audiences are different and therefore have different expectations and needs.

Consider our courtroom scenario again. Lawyers are courtroom strategists. They will base their case on the make-up of the jury and the judge presiding. When the jury has been selected, they will determine their backgrounds — are they middle class, upper class, etc? What is the ratio of men to women? What is the ethnic make-up of the jury? What is the judge's track record on cases similar to this one? What have the judge's previous rulings been based on? Take this same scenario and we begin to understand that learning about the people we must influence is a must.

In preparing for the final presentation, determine who the attendees will be. Then learn about their backgrounds. Are they technical people, financial people, or perhaps marketing and sales people? This will be of great value because making a technical presentation to a financial group would risk the success of the meeting.

Next we must determine what makes these people "tick." How are these people's incentives paid? Is it based on throughput, cost reduction, profitability, various ratios, or safety records? This becomes very important because when making the presentation, we must present the benefits of implementing the recommendations in units that appeal to the audience. For example, If we are able to correct this start-up procedure and provide the operators the appropriate training, based on past history, we will be able to increase throughput by 8% which will equate to $9.2 million in the first year!

HAVE AN AGENDA FOR THE MEETING

No matter what type of presentation that you have, always have an agenda prepared for such a formal presentation. Management typically expects this formality and its also shows organizational skills on the part of the team. Table 9.2 shows a sample agenda that we typically follow in our RCA presentations.

Always follow the agenda; only break away when requested by the management team. Notice that the last item on the agenda is "Commitment to Action." This is a very important agenda item as sometimes we tend to leave such meetings with a feeling of emptiness and we turn to our partner and ask, "How do you think it went?" Until this point we have done a great deal of work and we should not have to wonder how it went. It is not impolite or too forward at the conclusion of the meeting to ask, if it is unclear at that point, "Where do we go from here?" Even a decision to do nothing is a decision, and you know where you stand. Never leave the final meeting wondering how it went.

TABLE 9.2
Sample Final Presentation Agenda

#	Agenda Topic	Speaker
1	Review of PROACT™ Process	RJL
2	Summary of Undesirable Event	RJL
3	Description of Error Chain Found	KCL
4	Logic Tree Review	KCL
5	Root Cause Action Matrix Review	WTB
6	Recognition of Participants Involved	WCW
7	Question and Answer Session	ALL
8	Commitment to Action/Plan Development	RJL

DEVELOP A CLEAR, CONCISE, PROFESSIONAL PRESENTATION

Research shows that the average attention span of individuals in these managerial positions is about 20 to 30 minutes. The presentation portion of the meeting should be designed to accommodate this time frame. We recommend that the entire meeting last no more than one hour. The remaining time will be left to review recommendations and develop action plans.

The presentation should be molded around the agenda we developed earlier. Typical presentation software such as Microsoft's PowerPoint™* provides excellent graphic capabilities and also easily allows the integration of digital graphics, animation, and sound clips. Remember this is our day in court and we must bring out all the bells and whistles to sell our recommendations. The use of various forms of media during a presentation provides an interesting forum for the audience as well as aids in retention of the information by the students.

There is a complete psychology behind how the human mind tends to react to various colors. This type of information should be considered during presentation development. Overhead projectors, easel pads, 35-mm projectors, and better yet, the LCD projectors provide an array of different media to spice up the presentation. Props such as failed parts or pictures from the scene can be used to pass around the audience and enhance interest and retention. All of this increases the chances of the acceptance of recommendations.

Always dress the part for the presentation. Our rule of thumb has always been to dress one level above the audience. We do not want to appear too informal, but we also do not want to appear to stuffy. The appearance as well as the demeanor should always be professional. Remember we perform a $5 failure analysis on a $5 failure. This presentation is intended for a "significant few" item and the associated preparation should reflect its importance.

---

* PowerPoint® is a registered trademark of Microsoft.

Coordinate the Media to Use in the Presentation

As discussed earlier, many forms of media should be used to make the presentation. To that end, coordination of the use of these items should be worked out ahead of time to ensure proper "flow" of the presentation. This is very important, as a lack of such preparation could affect results of the meeting and show a disconnected or unorganized appearance.

Assignment of tasks should be made prior to the final presentation. Such assignments may include a person to display overheads while the other presents, a person to manipulate the computer while the other presents, a person to hand out materials or props at the speaker's request, and a person who will provide verification data at the request of management. Such preparation and organization really shines during a presentation and it is apparent to the audience.

It is also very important to understand the layout logistics of the room that you are presenting in. Nothing is worse than showing up at a conference room and realizing that your laptop is not compatible with the LCD projector. Then you spend valuable time fidgeting with trying to make it work in front of the audience. Some things to keep in mind to this end:

- Know how many will be in your audience and where they will be sitting.
- Use name cards if you wish to place certain people in certain positions in the audience.
- Ensure that everyone can see your presentation from where they are sitting.
- Ensure that you have enough handout material (if applicable).
- Prior to the meeting, ensure that your A/V equipment is fully functional.

Like everything else about RCA, we must be proactive in our preparation for the final presentation. After all, if we blow it in the courtroom, then our RCA cannot be successful as we will not have improved the bottomline.

Conduct Dry Runs of the Final Presentation

The final presentation should not be the testing grounds for the presentation. No matter how prepared we are, we must display some modesty and realize that there is a possibility that we may have holes in our presentation and our logic.

We are advocates of at least two dry runs of the presentation prior to the final. We also suggest that such dry runs be presented in front of the best constructive critics in your organization. Such people will be happy to identify logic holes, therefore strengthening the logic of the tree. The time to find gaps in logic is prior to the final presentation, not during. Logic holes that are found during the final presentation will ultimately damage the credibility of the entire logic tree. This is a key step in preparation for the final.

Quantify Effectiveness of Meeting

Earlier we discussed obtaining the recommendation acceptance criteria from management prior to developing recommendations. If these rules of the game are provided,

TABLE 9.3
Sample Quantitative Evaluation Form

Recommendation	Must Eliminate Cause	Must Provide a 20% ROI	Must Not Interfere with Any Capital Projects on Books	Average

then this offers a basis where we quantify our meeting results if our management is progressive enough to use quantification tools.

We recommend the use of an evaluation tool during the presentation of the recommendations that would require the management review group to evaluate each recommendation against its compliance with the predetermined recommendation acceptance criteria. Table 9.3 provides a sample cross section of such an evaluation tool.

If used, this form should be developed prior to the final meeting. Make as many copies as there are evaluators. As shown, the recommendations should be listed on the rows and the recommendation acceptance criteria should be listed across the columns. As we are making our presentation with regards to various recommendations, we would ask the evaluators to rate the recommendation against the criteria using a scale of 0 to 5. A 5 rating would indicate that the recommendation is on target and meets the criteria given by the management. A 0 on the other hand would indicate that the recommendation absolutely does not comply with the criteria set forth.

Based on the number of evaluators, we would take averages for how each recommendation fared against each criteria item and then take the average of those items for each recommendation and obtain a total average for how well each recommendation matched all criteria. Table 9.4 is a sample of a completed evaluation form.

Once this form has been completed, it can be applied to a predetermined scale such as the one shown in Table 9.5.

Once this process has been completed, we all understand what corrective action will be taken, which recommendations need modification, and which were rejected. This process allows interaction with the management during the presentation. It also allows for discussions that may arise when one manager rates a recommendation against a criteria with a 0 and another rates the same with a 5. Such disparities beg an explanation as to why the perspectives are so far apart. This is an unbiased and nonthreatening approach to quantifiably evaluating recommendations in the final presentation. It has been our experience, though, that only the very open-minded management would participate in such an activity.

PRIORITIZE RECOMMENDATIONS BASED ON IMPACT AND EFFORT

Part of getting what we want from such a presentation involves presenting the information in a "digestible" format. For instance, if you have completed an RCA

TABLE 9.4
Sample Completed Quantitative Evaluation Form

Recommendation	Must Eliminate Cause	Must Provide a 20% ROI	Must Not Interfere with Any Capital Projects on Books	Average
Modify maintenance procedures to enhance precision work	3.5	4	5	4.17
Design, implement, and instruct lubricators on how, when, and where to	3	5	2	3.33
Develop a 3-hour training program to educate lubricators on the science of tribology	4	5	4	4.33

TABLE 9.5
Sample Evaluation Scale

Average Score	Accept As Is	Accept with Modification	Reject
>= 3.75	X		
>= 2.5 < 3.75		X	
< 2.5			X

and have developed 59 recommendations, now the task is to get them completed. As we well know, if we put 59 recommendations on someone's desk, there is a reduced likelihood that any will get done. Therefore, we must present them in a "digestible" manner. We must present them in such a format that it does not seem as much as it really is. How do we accomplish this task?

We use what we call an impact-effort priority matrix. This is a simple three-by-three table with the X-axis indicating impact and the Y-axis indicating effort to complete. Figure 9.1 shows a sample of such a table.

Let's return to our previous scenario of having 59 recommendations. At this point we can say that we can separate the recommendations that we have direct control to execute and determine them to be high-impact, low-effort recommendations. Maybe we deem several other recommendations as requiring other departments' approvals, therefore they may be a little more difficult to implement. Finally, maybe we determine that some recommendations require that a shutdown occur before the corrective action can be taken. Therefore, they may be more difficult to implement. This is a subjective evaluation that breaks down a perceived impossible pile into manageable and accomplishable tasks. A completed matrix may look like Figure 9.2.

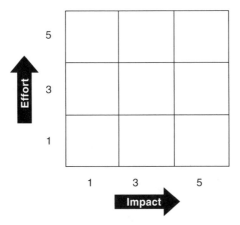

FIGURE 9.1 Impact-effort priority matrix.

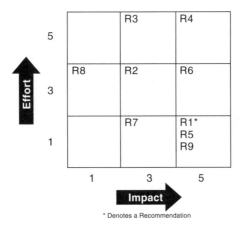

FIGURE 9.2 Completed impact-effort priority matrix.

DETERMINE NEXT STEP STRATEGY

The ultimate result we are looking for from this step (communicate findings and recommendations) is a correction action plan. This entire section is dealing with selling the recommendations and gaining approval to move on them. After the meeting we should have recommendations that have been approved, individuals assigned to execute them, and time lines to complete them by. The next phase to explore will be the effectiveness of implementation and overall impact on bottom-line performance.

10 Tracking for Results

Consider what we have accomplished this far in the RCA process:

1. Established management systems to support RCA
2. Conducted a modified FMEA
3. Developed a data preservation strategy
4. Organized an ideal RCA team
5. Used a disciplined method to draw accurate root causes
6. Prepared a formal RCA report and presentation for management

Up until this point, this has been an immense amount of work and an accomplishment in and of itself. However, success is not defined as identifying root causes and developing recommendations. Something has to improve as a result of implementing the recommendations!

Always remember that we are continually selling our need to survive, whether it is in society or in our organization. We must be constantly proving why we are more valuable than others to the facility. Tracking for results actually becomes our measurement of our success in our RCA effort. Therefore, since this is a reflection of our work, we should be diligent in measuring our progress because it will be viewed as a report card of sorts. Once we establish successes, we must exploit them by publicizing them for maximum personal and organization benefit. The more people that are aware and recognize the success of our efforts, the more they will view us as people to depend on in order to eliminate problems. This makes us a valuable resource to the organization. Make note that the more successful we are at RCA, the rewards should be that we get to do it again. This will be because the various departments or areas will start to request the RCA service from us. While this is a good indication, there can be drawbacks.

For instance, we have been trained to work on the "significant few" events that are costing 80% of the organization's losses. Under the described circumstances, we may have numerous people asking us to solve their pet problems that are not necessarily important to the organization as a whole. Therefore, when we decline, we may be viewed as not being a team player because we insisted on sticking to the "significant few" list from the modified FMEA. These are legitimate concerns we should have and discuss with our champions and drivers in how to overcome in our respective environments.

Let's pick up from the point where management has approved various recommendations of ours in our final meeting. Now what happens? We must consider each of the following steps:

1. Getting proactive work orders accomplished in a reactive backlog
2. Sliding the proactive work scale
3. Developing tracking metric(s)
4. Exploiting successes
5. Creating a critical mass
6. Recognizing the life cycle effects of RCA on the organization

GETTING PROACTIVE WORK ORDERS ACCOMPLISHED IN A REACTIVE BACKLOG

Unless approved recommendations are implemented, we certainly cannot expect phenomenal results. Therefore, we must be diligent in our efforts to push the approved recommendations all the way through the system. One roadblock that we have repeatedly run into is the fact that people generally perceive recommendations from RCAs as improvement work or proactive work. In the midst of a reactive backlog of work orders, a proactive one does not stand a chance.

Most computerized maintenance management systems (CMMS), or their industry equivalents, possess a feature by which work orders are prioritized by the originator. Naturally all originators think that their work is more important than anyone else's so they put the highest priority on the work order. Many work order systems' priority ranking system goes something like this (or equivalent):

1. E = Emergency — Respond Immediately
2. 1 = 24-Hour Response Required
3. 2 = 48-Hour Response Required
4. 3 = 1-Week Response Required

What normally happens with such prioritization systems is that 90% of the work orders are entered as "E" tickets requiring the original schedule to be broken in order to accommodate. Usually the preventive and predictive inspections are the items that are first to get bumped from the schedule, the proactive work!

Given this scenario, what priority would a recommendation from an RCA have? Typically, a 4! Such work is deemed back-burner work and it can wait because the event is not occurring now. This is an endless cycle if the chain is not broken. This is like waiting to fix the hole in the roof until it rains.

We mentioned earlier that management systems must be put into place to support RCA efforts. This is one system that must be in place prior to even beginning RCA. If the recommendations are never going to be executed, then the RCA should never begin. Accommodations must be made in the work order system to give proactive work a fair shake against the reactive work. This will require planners and schedulers to agree that a certain percentage of the maintenance resources must be allocated to executing the proactive work, no matter what. This is hard to do both in theory and practice. But the fact of the matter is that if we do not take measures to prevent recurrence of undesirable events, then we are acknowledging defeat against them and accepting reaction as the maintenance strategy. If we do not initially allocate some degree of resources to proactive work, then we will always be hung in a reactive cycle.

The answer to the above paradox can be quite simple. We have seen companies simply identify a designation for proactive work and ensure that the planners and schedulers treat them as if an "E" ticket with the resources they have set aside to address such opportunities. Maybe it's a "P" for proactive work or a block of worker numbers. Whatever the case may be, consideration must be given to making sure that proactive work orders generated from RCAs are implemented in the field.

SLIDING THE PROACTIVE WORK SCALE

As we hear all the time, the most common objection to performing RCA in the field is that we do not have the time. When we look at this objection introspectively, we find that we do not have the time because we are too busy "fighting fires." This truly is an oxymoron. RCA is designed to eliminate the need to fight fires. Management must realize this and include RCA as part of the overall maintenance strategy.

One way we have seen this done is through an interactive board game developed originally within DuPont™* and now licensed through a company in Kingwood, TX, called The Manufacturing Game©.** Organizational development experts within DuPont® developed this game. It is an innovative way in which to involve all perspectives of a manufacturing plant. When we played The Manufacturing Game© we found it to be an invaluable tool to show why a facility must allocate some initial resources to proactive work in order to remain competitive and in business. The Manufacturing Game© showed why proactive activities were needed to eliminate the need to do work and RCA expressed how to actually do it.

Proaction and reaction should be inversely proportional. The more proactive tasks performed, the less reactive work there should be. Therefore, all the personnel we currently have conducting strictly reactive work will now have more time to face the challenges of proactive work. We have yet to go to a facility that admittedly has all the resources they would like to conduct proactive tasks such as visual inspections, predictive maintenance, preventive maintenance, RCA, lubrication, etc. We do not have these resources now because they are all fire fighting. As the level of proaction increases, the level of reaction will decrease. This is a point where we gain control of the operation and the operation does not control us!

DEVELOPING TRACKING METRICS

Recognizing that inverse relationship between proaction and reaction, we must now focus on how to measure the effects of implemented recommendations. See Figure 10.1. This is generally not a complex task because typically there was an existing measurement system in place that identified a deficiency in the first place. By the time the RCA has been completed and the causes all identified, the metric to measure usually becomes obvious.

---

* DuPont is a registered trademark of the E. I. DuPont de Nemours & Co.
** ® 1998 The Manufacturing Game.

FIGURE 10.1 Inverse relationship between reaction and proaction.

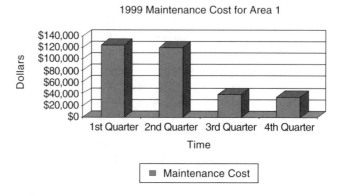

FIGURE 10.2 Mechanical tracking example.

Let's review a few circumstances to determine appropriate metrics:

1. *Mechanical* — We experience a mean time between failure (MTBF) of 3 months on a centrifugal pump. We find various causes that include a change of service within the past year, a new bearing manufacturer is being used, and the lubrication task has been shifted to operations personnel. We take corrective actions to properly size the pump for the new service, ensure that the new bearings are appropriate for the new service, and monitor that the lubrication tasks are being performed and in a timely manner. With all these changes, we now must measure their effectiveness on the bottom line. We knew we had an undesirable situation when the MTBF was 3 months; we should now measure the MTBF over the next year. If we are successful, then we should not incur any more failures during that time period due to the causes identified in the RCA. The bottom-line effect should be that savings are realized by man-hours not expended on repairing the pump, materials not used in repairing the pump, and downtime not lost due to lack of availability of the pump. See Figure 10.2.

2. *Operational* — We experience an excessive amount of rework (8%) due to production problems that result in poor quality product that cannot be sold to our customers. We find as a result of our RCA that we have

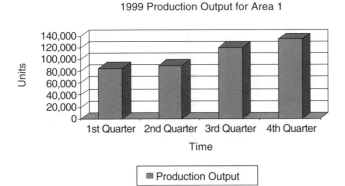

FIGURE 10.3 Operational tracking example.

instrumentation in the process that is not capable of handling a recent design modification. We also find that there are inconsistencies from shift to shift in the way the same process is operated. These inconsistencies are the result of not having written operating procedures. We implement the corrective action of installing instrumentation that will provide the information we require and writing a new operating procedure that ensures continuity. Rework started at 8%, so after we implement our solutions we should monitor this metric and make sure it comes down significantly. The bottom-line effect is that if we are reducing rework by 8%, we should be increasing saleable product by an equal amount while not incurring the costs associated with rework. See Figure 10.3.

3. *Customer Service* — We experience a customer complaint increase from 2 to 5% within a 3-month period. Upon conclusion of the RCA, we find that 80% of the complaints are due to late deliveries of our product to our clients' sites. Causes are determined to be a lack of communication between purchasing and the delivery firm on pick-up times and destination times. Also we find that the delivery firm needs a minimum of four hours notice to guarantee on-time delivery and we have been giving them only two hours notice on many occasions. As a result, we have a meeting between the purchasing personnel and the dispatch personnel from the delivery firm. A mutually agreed upon procedure is developed to weed out any miscommunications. Purchasing further agrees to honor their agreement with the delivery company of providing a minimum four-hour notice. Exceptions will be reviewed by the delivery firm, but cannot be guaranteed. The metric we could use to measure success will be the reduction in customer complaints due to late deliveries. See Figure 10.4.

4. *Safety* — We experience an unusually high number of incidents of back sprain in a package delivery hub. As a result of the RCA, we find causes such as lack of training in how to properly lift using the legs, lack of warming up the muscles to be used, and heavy-package trucks being assigned to those not experienced in proper lifting techniques. Corrective

1999 Customer Complaints for Area 1

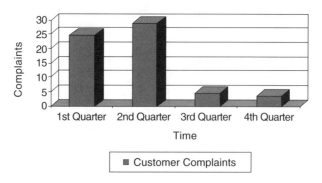

FIGURE 10.4 Customer service tracking example.

1999 Safety Incidents for Area 1

FIGURE 10.5 Safety tracking example.

actions include a mandatory warm-up period prior to the shift start, attendance at a mandatory training course on how to lift properly, passing of tests to demonstrate skills learned, and modifying truck assignments to ensure that experienced and qualified loaders/unloaders are assigned to more challenging loads. Metrics to measure can include the reduction in the number of monthly back sprain claims and also the reduction in insurance costs and workman's compensation to address the claims. See Figure 10.5.

The pattern of metric development described above shows that the metric that initially indicated that something was wrong can also be (and usually is) the same metric that can indicate that something is improving. Sometimes this phase seems too simple and therefore it cannot be. Then we start our "paralysis by analysis" paradigms and develop complex measurement techniques that can be overkill. Not to say they are never warranted, but we should be sure that we do not complicate issues that do not require it.

EXPLOITING SUCCESSES

If no one knows the successes generated from RCA, then the initiative will have a hard road ahead of it and the organization will not be optimizing the effects of the analyses. Like any new initiative in an organization, skepticism abounds at its survival chances. We discussed earlier the "program-of-the-month" mentality that is likely to set in after the introduction of such initiatives. To combat this hurdle, we need to exploit successes from RCA to improve the chances that the initiative will remain viable and accepted by the work population. Without this participation and acceptance, the effort is typically doomed.

One of the main ways we exploit such successes for our clients is through high-exposure mediums. High-exposure mediums include such media as report distribution, internal newsletters, corporate newsletters, company intranet, presentation of success at trade conferences, written articles for trade publications, and finally, exposures in texts such as this for successes demonstrated through the use of case histories. Exploitation serves a dual purpose — it gives recognition to the corporation as a progressive entity that uses its workforce's brainpower, and it provides the analyst and core team recognition for a job well done. This will be the motivator for continuing to perform such work. Without recognition, we tend to move on to other things because there is no glory in this type of work.

Let's explore the different media we just described:

1. *Report Distribution* — As discussed in the reporting section, to optimize the impact of RCA, the results must be communicated to the people that can best use the information. In the process of doing this, we are also communicating to these facilities that we are doing some pretty good work in the name of RCA and that our people are being recognized for it.
2. *Internal Newsletters* — Most corporations have some essence of a newsletter. These newsletters serve the same purpose as a newspaper, to communicate useful information to its readers. Most publishers of internal newsletters that we have ever dealt with would welcome such success stories for use in their newsletter. That is what the newsletter is for; therefore, we should take advantage of the opportunity.
3. *Corporate Newsletters* — Again, most corporations we deal with have some type of corporate newsletter. It may not be published as frequently as the internal newsletter, but nonetheless, it is published. These newsletters focus on the "big" picture when compared to the internal newsletter and may include more articles geared towards financials, overseas competition, etc. However, they too are looking for success stories that can demonstrate how to save the corporation money and to recognize sites that are exemplary.
4. *Presentations at Trade Conferences* — This is a great form of recognition for both the individual (and team) and the corporation. For some analysts, this is their first appearance in a public forum. While some may be hesitant at the public speaking aspect of the event, they are generally very impressed with their ability to get through it and receive the applause of

an appreciative crowd. They are also more prone to want to do it again in the future. Trade conferences thrive on the input of the companies involved in the conference. They are made up of such successes and the conference is a forum to communicate the valuable information to others that can learn from it.

5. *Articles in Trade Publications* — As we continue along these various forms of media, the exposures become more widespread. When we get to talking about trade publications we are now talking about exposure to hundreds of thousands of individuals in the circulation of the magazine. The reprints of these articles tend to be viewed as trophies to the analysts who are not used to such recognition. As a matter of fact, when we have such star client analysts who have written an article about their success, we frame the reprint and send it to the analyst for display in the office. It is something they should be proud of as an accomplishment in their career.

6. *Case Histories in Technical Text* — As we will read in the remainder of this text, we solicited responses from our client base on interested corporations that would like to let the general public know of the progressive work they are doing in the area of RCA and how their workforce is making an impact on the bottom line. As most any corporation will attest, no matter what the initiative is or what the new technology may be, without a complete understanding of how to use the new information and its benefits by the workforce (personally and for the corporation), it likely will not succeed. Buy-in and acceptance produce results, not intentions or expectations of the corporation.

CREATING A CRITICAL MASS

When discussing the term *critical mass,* we are referring not only to RCA efforts, but the introduction of any new technology. It has been our firm's experience in training and implementation of RCA efforts over the past 26 years, that if we can create a critical mass of 30% of the people on board, the others will follow.

We have beat to death the program-of-the-month mentality, but it is reality. Some people are leaders and others are followers. The leaders are generally the risk takers and the ones who welcome new technologies to try out. The followers are typically more conservative people who take the "let's wait and see" attitude. They believe that if this is another program-of-the-month, they will wait it out to see if it has any staying power. These individuals are those who have been hyped up before about such new efforts, possibly even participated, and then never heard any feedback about their work. They are in essence alienated with regards to "new" thinking and the seriousness of management to support it.

We believe that if we can get 30% of the population trained in RCA to actually use the new skill in the field and produce bottom-line results, then RCA will become more institutionalized in the organization. If only 30% of the analysts start to show financial results, the dollars saved will be phenomenal; phenomenal enough to catch executives' eyes where they continue to support the effort with actions, not words.

Once the analysts start to get recognition within the organization and corporation, others will crave similar recognition and start to participate.

We have found it unrealistic to expect that everyone we train will respond in the manner that we (and the organization) would like. It is realistic to expect a certain percentage of the population to take the new skills to heart and produce results that will encourage others to come on board.

RECOGNIZING THE LIFE CYCLE EFFECTS OF RCA ON THE ORGANIZATION

RCA can play a major role in today's overall corporate strategies for growth. As we have referred to throughout this text, the goal should be the elimination of the recurrence of any undesirable outcomes that have occurred in the past. Many organizations set their sights and hence their standards at being the best "predictors" of such events and thus are targeting the reduction in response time as the successful measure. While this is still a must in the interim, it should be a means to another end — the elimination of the recurrence. If we did not have undesirable outcomes, we would not have a need to become better predictors.

We have seen millions and millions of dollars spent by corporations around the world on reliability-centered maintenance (RCM). In its textbook implementation, RCM is ultimately geared towards helping firms determine the criticality of equipment and processes in their operations and then develop a customized preventive and predictive maintenance strategy based on that information. The end result is that we have a very in-depth understanding of our operation and what could ever possibly go wrong. Most of those corporations that have taken to heart RCM will agree that it is very expensive to implement and extremely resource consuming. However, the yields from such efforts are typically incremental in the short term.

While we have seen many organizations grasp the RCM concept, we continue to have difficulty in convincing corporations to give equal credence to an RCA effort. When implemented appropriately, RCA is eliminating the recurrence of events that are occurring now, and they are even being compensated for in the budget. When such "chronic" issues are solved and eliminated, there is no need to budget for their occurrence any more. The savings are off the bottom line in the same fiscal year.

We are in agreement with the concept of RCM in general; however, much time can be spent analyzing how to combat an event that has a miniscule chance of ever occurring. RCA is geared towards working on events that have and are occurring. RCM and RCA are complementary efforts toward total elimination of undesirable events. Over the past decades we have been inundated with what we call the *zero imperatives*. The zero imperatives are the efforts associated with zero touch labor, zero inventory, zero injuries, zero quality defects, etc. RCA is geared towards zero failures, or the elimination of undesirable events. See Figure 10.6.

While we are all realistic about these zero imperatives, we realize that they are literally not obtainable but they do provide the point to strive for. If our stockholders had their druthers, they would want the assets in any facility to run 24 hours a day for 365 days a year at maximum capacity in a sold-out market. This will never happen without a zero failure environment!

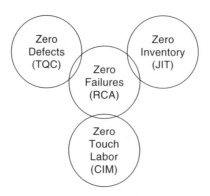

FIGURE 10.6 The zero imperatives.

CONCLUSION

Let's face the facts, we are a human species and we are evolving. We may never be perfect, but that should not preclude us from striving to be so. We will never be error-free, but we can strive to be. Precision is a state of mind and requires the mentality to constantly strive for the next plateau.

RCA as described in this text is not a panacea. It is merely a method to assist in logical thinking to resolve undesirable events. While many of our analogies have been from the industrial world where our background lies, we hope that it is clear that this RCA approach is applicable under any circumstances. Whether it is chronic or sporadic, mechanical or administrative, or in an oil refinery or a hospital, they all require the same logical human thought process to resolve their respective issues.

In the following chapters we will discuss how to make this thought process more manageable. We will seek to alleviate the administrative burden of managing an RCA by providing a simple and effective software solution. While conducting RCA in a disciplined manner as we have preached in this text can be difficult, most of the time is spent in sticking to the discipline and documenting the process. One of the ways we can provide an incentive to take this extra step of "discipline" is to make the task easier and more desirable. This is where the PROACT® software plays its role.

Finally, we will show the field proven benefits received by those firms who had the courage to adhere to the PROACT® discipline and produce phenomenal results for their companies.

11 Automating Root Cause Analysis: The Utilization of PROACT® Version 2.0

PROACT®* is the acronym that we have been using throughout this text to describe a root cause analysis (RCA) method. In this chapter we will use PROACT® (Version 2.0) to describe the name of a software package that facilities an RCA. In this chapter we will relate how and where there are opportunities to automate tasks that are otherwise done manually in the performance of an RCA.

From the standpoint of where do we start, we have discussed at length the pros and cons of conducting modified failure modes and effects analysis (FMEA), both manual and automated, using various sources of data. Data sources generally come from current computerized maintenance management systems (CMMS) institutionalized within organizations. Reliance on this data can be a good point to get started; however, it must be realized that "sleepers" exist. Sleepers are the tasks that happen so often that they are typically not recorded in the CMMS. This is because these sleepers characteristically do not take a great deal of time to repair or correct. It is seen as a burden and a waste of time to enter it into the system because the entry could take longer than the fix. The problem comes when such sleepers happen 500 times a year and no recording mechanism picks them up. This is generally what the maintenance budget is picking up and accepting as a cost of doing business.

With all this said, we just want to make sure that the CMMS is not viewed as the cure-all to all RCA problems. Until we are 100% confident that the CMMS is truly reflective of the activity in the field, we should consider the use of interviews with the people that do the work as the greatest source of data for an FMEA.

CUSTOMIZING PROACT® FOR OUR FACILITY

One important feature of RCA software is that it should be customizable for our specific facility. This means that we would like to see accommodations for our site information (facility locations, divisions, and departments) and for our equipment listings to be input (type and class). This makes it much easier when we are working on specific analyses to be able to choose from a pick list of items that are familiar to us, and not simply picking categories that apply to any industry.

---

* PROACT® is a registered trademark of Reliability Center, Inc.

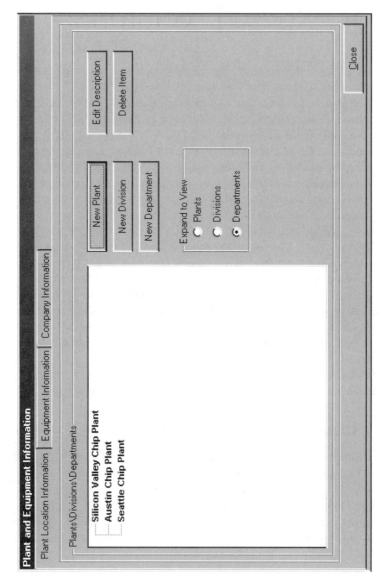

FIGURE 11.1 Plant and equipment information screen.

PROACT® allows for such information to be input to its databases for storage and retrieval. Figure 11.1 shows the ease in which we can manually input such information if it is not readily available, or we can take existing files with such information and import them into PROACT®.

When completed, PROACT® will have stored all of the various plant sites, their respective divisions and departments, and all the common equipment types and classes.

SETTING UP A NEW ANALYSIS

Once we have the plant and equipment information stored, we can set up a new analysis for us to start. PROACT® will carry users through a set-up wizard that will ask them to input certain base information about the analysis. This base information is stored in a database as well for use with search features to be described later on.

The new analysis wizard is a series of six steps. Step 1 is the inputting of the new analysis name, description, and type. The type is very important. The program permits the user to pick the analysis type from a pick list, citing such choices as safety, mechanical, environmental, operational, quality, and administrative. This will allow the eventual sorting of the analysis database on these categories. Therefore, when quality engineers want to view all of the completed analyses on quality issues, they can simply sort the database on this field. See Figure 11.2.

Step 2 will involve using the information we input above about our plant and equipment. Step 2 involves identifying the specifics about the event being analyzed. This will assist us later when trying to data mine a database for information about specific events that have been analyzed. See Figure 11.3.

Step 3 will involve the setting up of team members for the specific analysis at hand. Again, a prepopulated database will exist with our company's personnel. We will then be able to pick and choose who would be most suitable for this analysis (Figure 11.4).

Step 3 will also allow the principal analyst the latitude to grant permissions to each team member. Permissions include the following: (1) read only, (2) read/write, and (3) delete. This comes in handy when we may have temporary or contract employees on our team and do not wish to grant full permissions.

Step 4 will involve the setting up of the team information. This information includes when the analysis will start, when it is anticipated to be completed, and a free-form field for any general comments. See Figure 11.5.

Step 5 involves the setting up of specific team information (Figure 11.6). As discussed in Chapter 7, it is helpful to plan a direction or focus for the team. We talked about the team's charter and critical success factors (CSF). This is where that information would reside.

Step 6 is merely a confirmation that the analysis has been created successfully and that now we are ready to enter into the analysis itself. See Figure 11.7.

We will now venture into the "PR" of PROACT® and learn how to automate the data collection tasks.

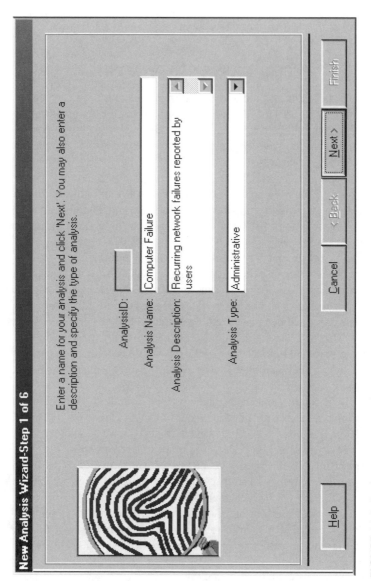

FIGURE 11.2 Setting up a new analysis, step 1.

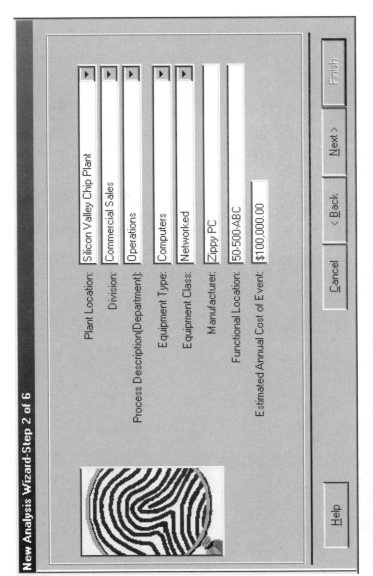

FIGURE 11.3 Setting up a new analysis, step 2.

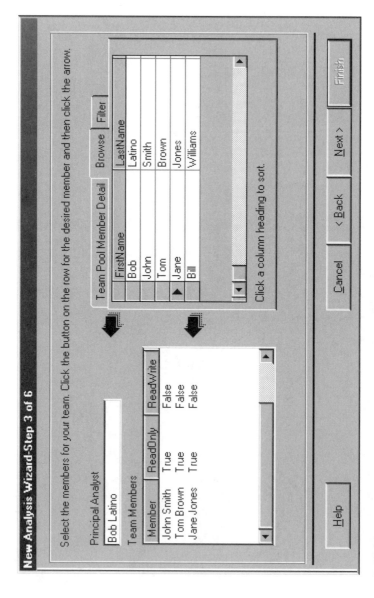

FIGURE 11.4 Setting up a new analysis, step 3.

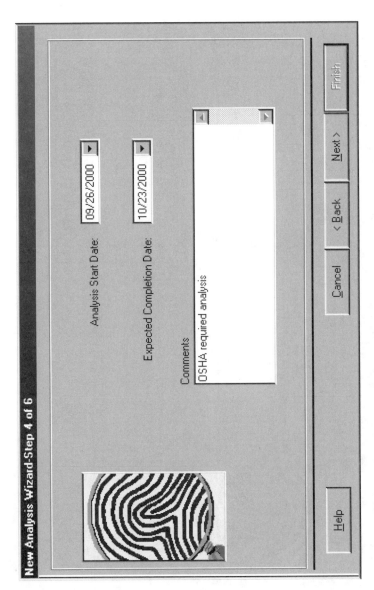

FIGURE 11.5 Setting up a new analysis, step 4.

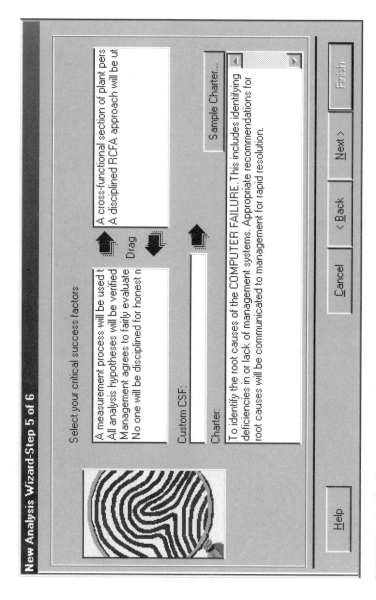

FIGURE 11.6 Setting up a new analysis, step 5.

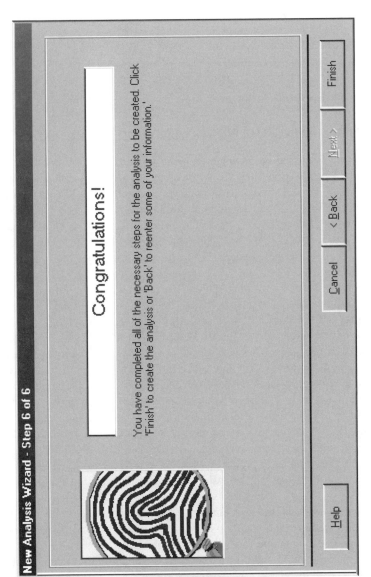

New Analysis Wizard - Step 6 of 6

Congratulations!

You have completed all of the necessary steps for the analysis to be created. Click 'Finish' to create the analysis or 'Back' to reenter some of your information.'

Help   Cancel   < Back   Next >   Finish

FIGURE 11.7 Setting up a new analysis, step 6.

Data Type: _____   Responsible: _____

Completion Date:__/__/__

Person to Interview/Item to be Collected: _____
_____
_____

Data Collection Strategy: _____
_____
_____
_____

FIGURE 11.8 Manual data collection form.

AUTOMATING THE PRESERVATION OF EVENT DATA

In Chapter 6 we discussed the manual approach to preserving event data utilizing our 5-Ps data collection strategy forms. While effective, it can lack efficiency because of the organizational skills required to manage the paperwork. Also from an efficiency standpoint, manual methods require double handling of data, which is non-value-added work. Whenever we write down information, it will eventually have to be reentered into a computer for final presentation. Automation provides an opportunity to eliminate these inefficiencies. To refresh our memories, the manual form looked like Figure 11.8.

So now that we know what information is required and in the format we desire it, the automation requirements have been determined. The screen shot in Figure 11.9 is from PROACT® and shows the tabs associated with the acronym.

Now we will open the tab designated PRESERVE and we find a table with the same fields as on the manual form, except they are now allowing a database to be developed in the background. See Figure 11.10.

The first team meeting, as described in Chapter 7, would involve a brainstorming session of the core team. The team would assemble and, based on the given facts at hand, start to develop a list of data that will be necessary to collect in order to start the analysis. This type of automation is most effective if a laptop is available in the meeting with an operator/recorder entering data as it is offered. The ideal situation is the use of an LCD projector with the laptop so that the entries are seen on the screen and everyone can be assured that their information was transcribed accurately.

As the type of data to collect is entered, a team member should be assigned to obtain that information using a certain collection strategy. A time frame should be assigned to focus the team and forge a progression of the analysis.

In the preserve module, the user is provided the opportunity to document their data collection strategy. PROACT® permits the user to link imported files to the desired preserve record. This allows the user to document the record with pictures of failed parts, results of lab testing, applicable procedures, and the like. This detail in documentation helps build our "solid case" with hard proof to back up our analysis. See Figure 11.11.

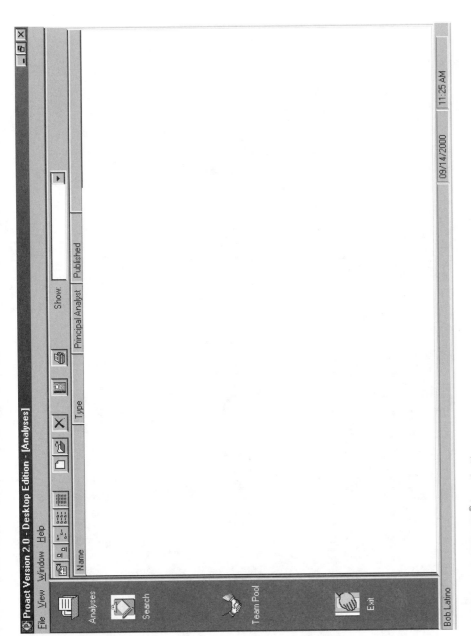

FIGURE 11.9 PROACT® introduction screen.

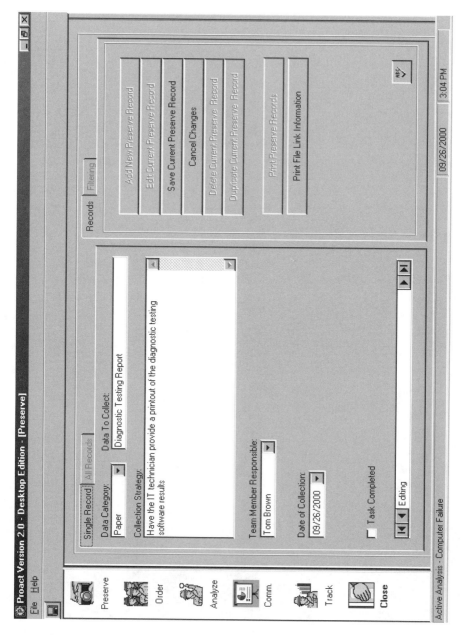

FIGURE 11.10 PRESERVE — opening screen.

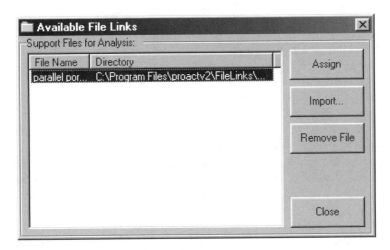

FIGURE 11.11 Preserve — file linking screen.

PROACT® provides a filtering and sorting option that will allow the principal analyst to print the lists at the end of the meeting in an array of formats. Sorts can be performed on various fields to make understanding easier for the recipients. Filters can be established to print only certain information and exclude all others. For instance, we may want to only print one team member's assignments vs. all team members. A browse feature is also available to see the entire list at a glance.

When using such automation, especially in a team format, it tends to maintain interest and, more important, it maintains organization of the entire RCA as we go through the entire process.

AUTOMATING THE ANALYSIS TEAM STRUCTURE

We discussed at length the importance of the team structure. If we reflect, we discussed how important the diversity of backgrounds was to a successful result. We also stressed that the leader of an RCA team should typically not be the expert in the event being analyzed because of the inherent bias that may persist.

We discussed the focus of the team structure by formalizing the team entity through the development of a team charter and the identification of critical success factors (CSFs). These tasks show management that there was considerable thought about why the team was formed and what their objectives are in obtaining success.

Now let's contrast the manual version with the automated version. In a manual format we would most likely be utilizing a paper filing system to record team member information. We would also likely use a word processing program to develop the team charter and the CSFs. With an automated format, we can use PROACT® to catalog all this information in one location along with the 5-Ps information collected previously.

Remember, all of this was collected during the initial new analysis wizard (six steps) when we created the new analysis. The Order the analysis team tab is merely where that information is stored and available for modifications. In the example below we show a change in granting permissions to a team member. This task is easily completed by the principal analyst by simply double-clicking on the team member's name. See Figure 11.12.

PROACT® will maintain a team pool by which a database of qualified RCA team members is stored. "Qualified team" members may be past RCA participants, individuals who have received RCA training in the past, or individuals who possess a certain expertise that is difficult to find. In any case, maintaining a record of such talent is an efficient way of helping to organize RCA teams. Once a reservoir of talent has been identified, then specific individuals can be assigned to lead and participate on the core team. These choices will obviously vary based on the nature of the event being analyzed. PROACT® will allow reports to be developed on the team members, based on their names and/or telephone numbers. See Figure 11.13.

PROACT® now has all the team information cataloged and organized within a database. Up until this point, there has been no need to use individual database or spreadsheet programs, or word processing programs. It is all located within one RCA file.

AUTOMATING THE ROOT CAUSE ANALYSIS — LOGIC TREE DEVELOPMENT

Moving on, let's assume we are at a point in the analysis where the initial data has been assigned and collected, and that an ideal team has been put together and organized to approach the task of analysis. Now we face the real issue of analyzing the data to determine what happened.

Using the manual method to develop the logic tree has its pros and cons. One of the disadvantages is that it is double handling of data. In the manual method, a logic tree is being built in a conference room where a mural has been put together made of easel pad paper or craft paper. Subsequently, the analysts will be facilitating the team using Post-Its®.* This means that at some point in time, this information will have to be transcribed in another format for inclusion in the report and/or display in a presentation. This double handling leads to an inefficiency of time as well. When a team meeting ends, the team members usually do not have the updated logic tree until days later. This results in unnecessary delays before all team members have consistent information.

One of the psychological advantages of using the manual method in conjunction with the automated method is that it can be perceived as accomplishing work. We have seen the paradigms at play where many believe that if someone is working on a computer all the time that work is not being accomplished. Some may feel that if wrenches are not being turned or machines are not being operated, then work is not being accomplished. The same can be said for RCA. If management walks by a conference room where an RCA team is meeting and only sees one laptop on the table and five team members sitting around talking, then it can be perceived as a non-value-added use of time. However, in the same scenario, if they walk by and see this huge craft paper on the wall with all these Post-Its®, then that can be deemed

---

* Post-It ® is a registered trademark of the 3M Company.

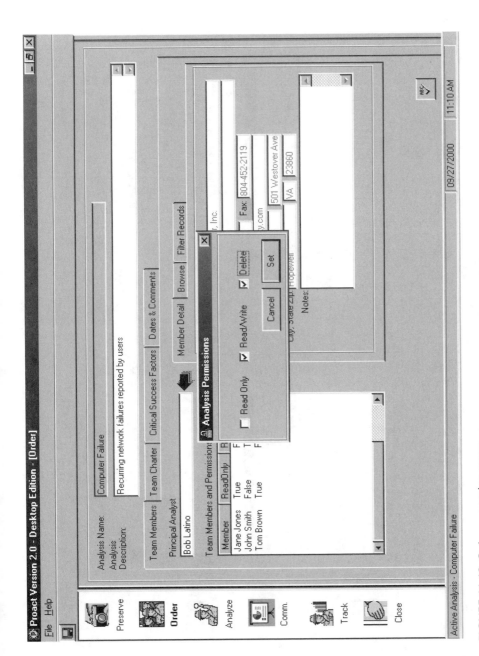

FIGURE 11.12 Order — opening screen.

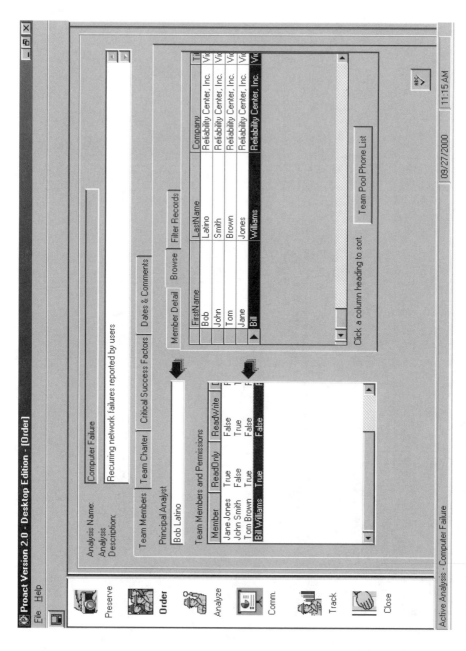

FIGURE 11.13 Order — team pool tab.

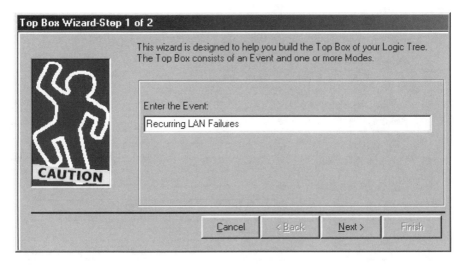

FIGURE 11.14 The Top Box Wizard — step 1.

as tangible work (even if a recorder has duplicated the logic tree within PROACT® on the laptop within the same meeting).

From an efficiency standpoint, using a laptop and an LCD projector in a team meeting is the ideal forum to conduct logic-tree-building sessions. This will obviously have to be the determination of the analyst or the team based on the resources they have available to them at their site.

We will now go through how PROACT® can help automate the analysis of data. PROACT® was developed using the same logic rules as discussed in the RCA method described in this text.

The opening screen in Analyze is basically a blank worksheet with the necessary tools or tabs to build the logic tree. When beginning to build the logic tree during a team meeting, the analyst should start with the Top Box Wizard tab. This will prompt the team to enter what exactly is the event that they are ultimately analyzing. The detailed descriptions of how to develop events and modes are located in Chapters 4 and 8. Once an event has been entered, the team will be prompted to enter the various modes that apply under the circumstances. Only enter as many modes as are necessary for the particular event being analyzed. See Figures 11.14, 11.15, and 11.16.

Visualize as we go through these scenarios that the LCD projection is on a screen and the entire team can view the tree-building as it is developed. At the same time, it is being recorded in the RCA file. If the top box has been outlined, then the known facts of the situation have been identified and now we must begin the process of hypothesizing how these facts could have occurred. Now this is where the analyst plays the role of a facilitator and begins the continual questioning process of how could the event have occurred. The core team of experts will be the source for the answers from various perspectives. As appropriate hypotheses are thrown on the table, they are entered into the logic tree. As each hypothesis is entered, the user

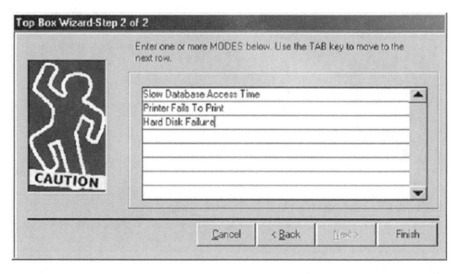

FIGURE 11.15 The Top Box Wizard — step 2.

will be reminded of the need to fill out the verification log. The verification log form will pop up every time a hypothesis is entered.

The verification log will:

1. Prompt the team to designate who will be responsible for testing the hypotheses.
2. List what test is to be performed.
3. List the anticipated completion date of the test.
4. List the actual completion date of the test.
5. List the test results and the confidence level that the hypothesis is true or false based on the test.

See Figure 11.17.

Once a verification log has been assigned a responsible person, a verification method or test to be done, and a completion date, then it is logged in and stored awaiting an outcome. Sometimes such verifications will require the attachment (file linking) of the proof of the verification test. This can be in the form of test results, pictures, reports, procedures, etc. As we see in Figure 11.18, this is easily accomplished (just as we did in Preserve).

As the reiterative process moves on and more and more hypotheses are developing, the logic tree continues to grow. The Tree Objects tab allows the users to drag and drop icons that indicate the block being developed. When Physical Roots are reached, the user simply drags the icon on top of the hypothesis and it is now signified as a physical root. Consequently, in the background of the program, this root cause will now automatically come up in the report writing section and require a recommendation be made to eliminate its recurrence. See Figure 11.19.

PROACT® also provides several features that allow the user to manipulate the program with ease. In the Tree Option tab, you will notice that the user has the

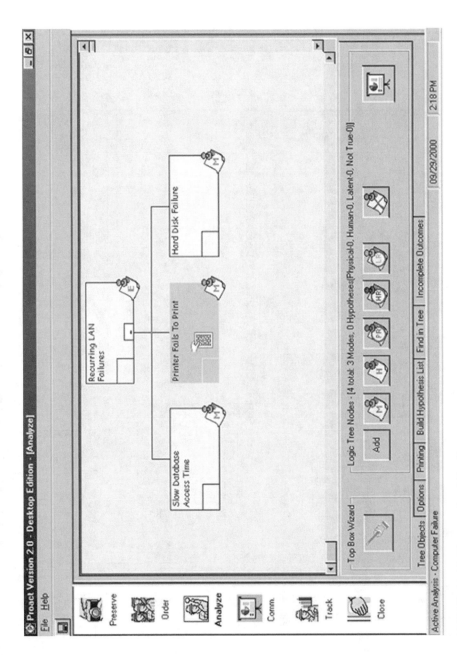

FIGURE 11.16 The completed Top Box.

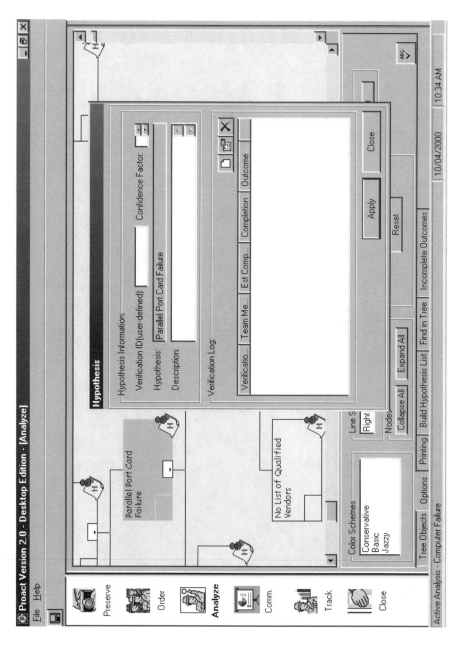

FIGURE 11.17 Analyze verification log development.

FIGURE 11.18 Verification log — file linking.

ability to search for various hypotheses based on their words or associated block numbers. The user can also expand and collapse the tree at will and move branches from one area to another via arrows. Zoom and Spacing options are also available to make viewing easier for certain individuals. Should the analyst wish to print the "to date" Verification Log Report, he can do so in this section. This means at the end of any given team meeting, the current log can be distributed to the team immediately. See Figure 11.20.

PROACT® provides the analyst many ways in which to print the team's logic tree. PROACT® will also print to ANSI A through ANSI E sized plotter paper allowing printing on roll-fed plotters. This makes a big difference when printing a logic tree with 100 and 200 blocks. See Figure 11.21.

At this point in the RCA, consideration will begin with regards to the final presentation of the logic tree to management. To this end, PROACT® provides a presentation mode to eliminate the need of developing a presentation using a graphics program in another separate file. By clicking the Screen tab, a full screen presentation mode will appear with the entire logic tree expanded. This mode will allow the team to make their final presentation real time. The speaker can begin with the collapsed logic tree showing only the event and modes. Then as the presentation progresses, the team can expand on the hypothesis in question and show the possibilities. If a manager questions any

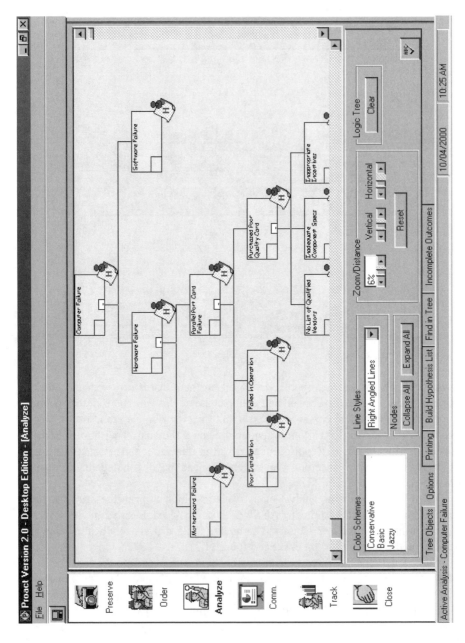

FIGURE 11.19 Logic tree with root causes.

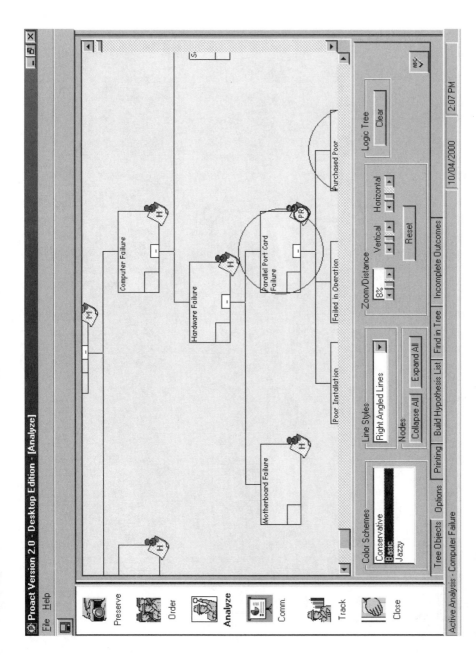

FIGURE 11.20 Logic tree — options.

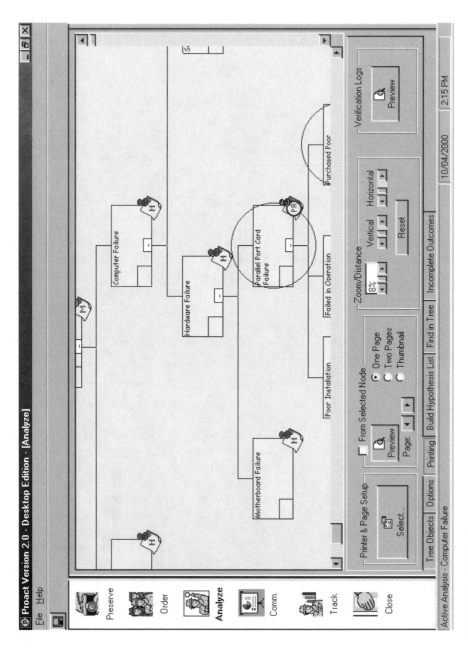

FIGURE 11.21 Logic tree — printing tab.

hypothesis, then all the speaker has to do is double-click on that block and the verification log pops up to show how the hypothesis was tested and what the result was. This is an extremely useful feature when making such a presentation. See Figure 11.22.

Considering the time it takes to develop a formal presentation and the subsequent supporting data such as verification information, PROACT® considerably reduces the time to perform such tasks. Again, thus far, all information that is related to this RCA is located in one file using one program.

AUTOMATING RCA REPORT WRITING

One of the most tedious tasks about conducting a full-blown RCA is the writing of the report. If no standard formats are available, this can be a laborious task that lacks continuity. Without standard formats, consistency of reporting results suffers and the information is ignored or not understood. In the manual method of writing reports, we would generally use a word processing program and develop a stand-alone report with a table of contents that suits the team. Then some poor soul, usually the principal analyst, is charged with the task of developing the content and typing it into an acceptable format. While the team members may contribute, the brunt of the legwork is on the shoulders of the principal analyst. Then the task of properly distributing the report to the parties that would benefit the most is at hand. All in all, the task is extremely burdensome and not the highlight of the analysis work.

PROACT® provides analysts with a report writer where the author must only fill in the blanks, which represent report topics. The customized report feature breaks the report into three sections:

1. The Executive Summary — Short Recommendations
2. The Detailed Section — Long Recommendations
3. The Appendices — Supporting Data

Each of these sections was discussed at length in Chapter 9. Our purpose here is to show how we can automate the report-writing task. Within the Executive Summary fields we are prompted to fill in an event summary, event mechanism, and PROACT® description. See Figure 11.23.

As we also discussed in Chapter 9, the entire RCA process revolves around the final presentation and getting recommendations approved. The root cause action matrix was the culmination of the entire analysis. To this end, PROACT® provides such a matrix, which requires input in various fields. Any hypothesis on the logic tree that was designated as a root cause in the Analyze section will automatically appear in the drop box along with the appropriate type of root cause it was identified as (physical, human, or latent). PROACT® will also seek a person on the team (or someone else) as being responsible for implementing the recommendation by a certain date. Therefore, when the logic tree has been completed, the roots, which require recommendations, should be assigned to various individuals and they should set target dates to complete them by. Remember, in this section we are merely giving a summary of the recommendation (short recommendation), which is designed to greatly reduce the impact of, or eliminate the identified cause.

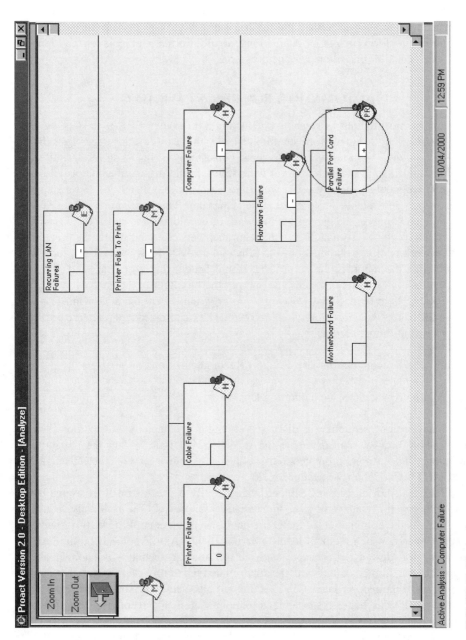

FIGURE 11.22 Analyze — presentation mode.

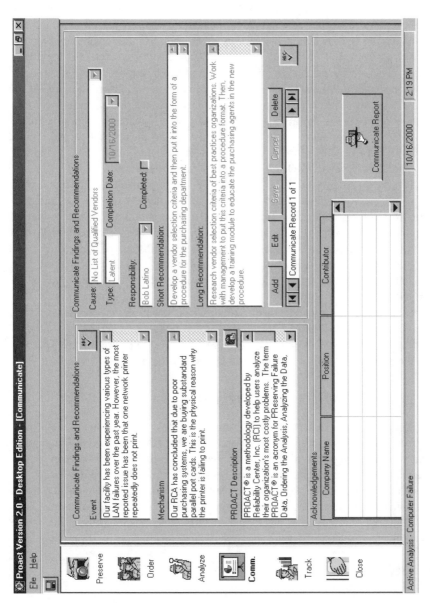

FIGURE 11.23 Communicate — executive summary (short recommendation) screen.

The second major section of the report is the Detailed Recommendations. This is very similar to the Executive Summary, however depth of recommendation is the difference here. Here is where the entire plan to execute and resource the recommendation is laid out with any cost/benefit analysis results. As in the Executive Summary, the root causes identified will automatically be accessible and the author will be required to input a detailed recommendation, or Long Recommendation.

Finally, the Acknowledgments section is set aside to give recognition to those who were instrumental in the success of the RCA effort other than the core team members. This is where that recognition is designed to take place. See Figure 11.24.

Of course, all of this effort is for naught if you cannot print the report. PROACT® allows the author to print the entire report, or to select sections of preference only. A Print Wizard allows the author to customize covers and also to print selected topics if desired. This automating of the report writing means that formal reports do not have to be developed from scratch; we do not have to worry about formatting or standardizing because PROACT® is doing it all automatically.

When developing Version 2.0 of PROACT®, almost all of the feedback from Version 1.0 users was incorporated. One of those changes was in the reporting section. Many users understood that PROACT was an acronym, but they preferred the option to print in any order they desired. As a result we added a Print Wizard feature that allowed them to do just that, customize their table of contents.

By unchecking the Use Default Topics, we can drag and drop the sections and subsections that we choose to include in this printing of the report. See Figure 11.25.

The next step in the Print Wizard (Figure 11.26) allows the user to put a title, subtitles, and a company logo on the report front cover if they desire. Also the desired printer information can be adjusted at this point.

The final step of the Print Wizard (Figure 11.27) is the print preview. This will allow the user to preview and scroll through the entire report.

AUTOMATING TRACKING METRICS

As we know, we are not successful at RCA unless some bottom-line metric improves. Therefore, we must select and monitor over time the metric of choice. In a manual format we may have to be diligent about getting certain data from certain reports or we may have to develop a whole new report to get the information we seek.

One thing we should not do is make the tracking process so complicated that it is too difficult and frustrating to accomplish. PROACT® was designed to make this tracking process very simple, basic, and user-friendly. Tracking also has its own four-step wizard, which will walk the user through a series of questions such as:

1. Save Graph as:_____
2. Title of Graph:_____
3. Sub-Title:_____
4. Tracking Intervals:_____
5. Tracking Periods:_____
6. Tracking Metric:_____
7. Data to Input:_____

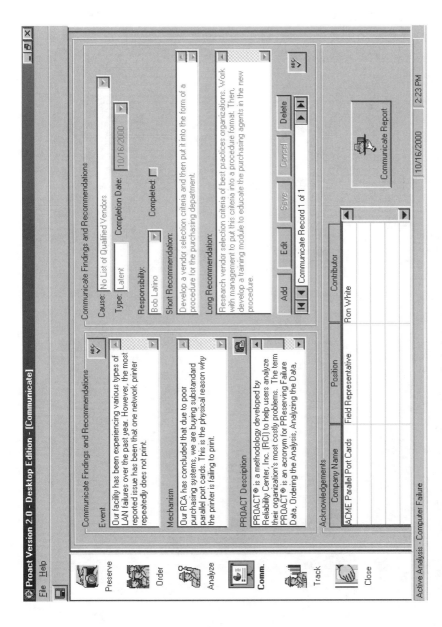

FIGURE 11.24 Communicate — long recommendations.

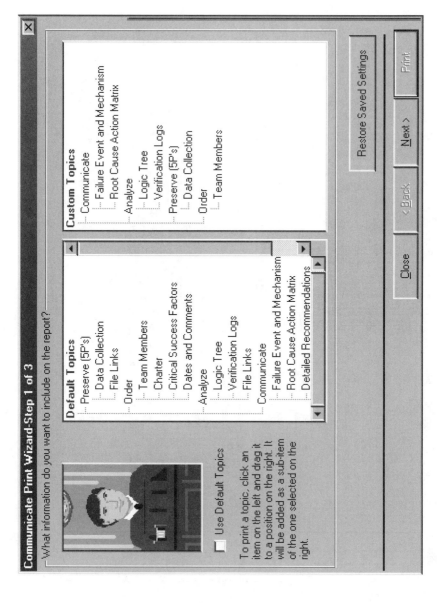

FIGURE 11.25 Communicate — print wizard step 1.

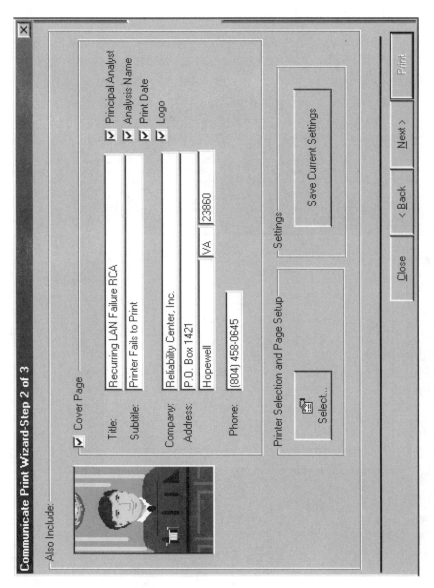

FIGURE 11.26 Communicate — print wizard step 2.

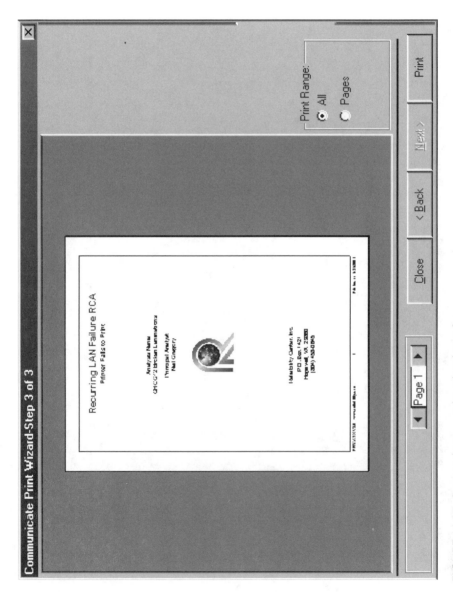

FIGURE 11.27 Communicate — print wizard step 3.

This provides enough data to make an easy-to-follow basic graph. See Figure 11.28. Each month when new data is available, it can be input into the wizard to update the graph rather easily. If using PROACT® in an enterprise environment that is integrated with other data systems such as Meridium's Enterprise Reliability Management System (ERMS)*, this data could be collected automatically.

If analysts choose to use more elaborate charting functions, they can easily export the chart to Excel and maintain a file link to its location. The development of a dynamic tracking graph completes the circle of finalizing an RCA. Automating this graphing feature in PROACT® alleviates the need to use a separate graphics package to make the graph.

Think back now, if we use the traditional manual method, we would require the use of a database package, a spreadsheet package, a word processing package, and a graphics package in order to complete the RCA. This would require the alignment of file names and so on for continuity. PROACT® compiles everything in one location and the file can be e-mailed to others to ensure proper distribution.

PROACT®'s enterprise version allows for the efficient and effective knowledge transfer of the successful analyses to others in the company that may benefit. PROACT® puts information at the fingertips of those who can use it most. Until an analysis is completed, only the principal analyst and the team members can see their work in progress. However, when they are done and the principal analyst certifies the analysis as complete, everyone that has permission and access to PROACT® will be able to view the results. This feature is called *publishing* (Figure 11.29). Once an analysis has been published, its icon within the database changes, allowing us to visually recognize those analyses completed. Also, *only* completed analyses can be searched based on various criteria.

These features, as well as many others, allow analysts to focus more of their time on doing the analysis rather than on the administrative tasks to document the process and transfer the knowledge. PROACT® is truly a proactive tool when conducting RCA!

It has been our experience that using PROACT® to facilitate RCAs in the field has reduced the administrative time to complete them by approximately 50%. This means from a productivity standpoint that analysts can complete more analyses in a given time period if they automate their RCA.

PROACT® was presented a Gold Medal Award (general maintenance software) in *Plant Engineering's* "Product of the Year" competition in 1999 and 2001 for both Versions 1 and 2, respectively. For more information about how to obtain PROACT® contact:

Reliability Center Incorporated
501 Westover Avenue, Suite #100
P.O. Box 1421
Hopewell, VA 23860
Telephone: 804-458-0645
Fax: 8040452-2119
Web address: http://www.reliability.com

---

* ERMS is a registered trademark of Meridium, Inc., Roanoke, VA.

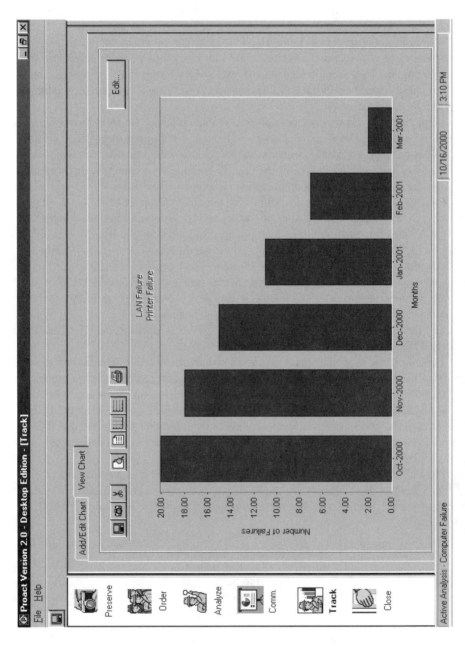

FIGURE 11.28 The tracking wizard — final chart.

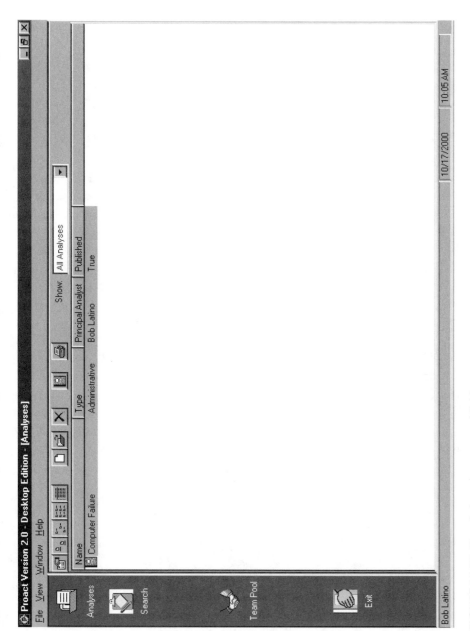

FIGURE 11.29 Knowledge manager — publishing.

12 Case Histories

This chapter will put into practice what this text has described in theory thus far. We have described in detail the RCA method and provided some academic examples to further understanding of the concepts.

The following case studies are a result of having the right combination of management support, ideal RCA team, and proper application of the RCA methodology. RCI commends the submitters of these case histories for their courage in allowing others to learn from their experiences. These corporations and their RCA efforts have proven what a well-focused organization can accomplish with the creative and innovative minds of their workforce.

As we read through the summaries of these actual case histories, we will notice that the returns-on-investment (ROI) for eliminating these chronic events range from 3,100% to 17,900%. Had we not had permission to publish these remarkable returns, would anyone have believed they were real? We will also notice that the time frames to complete the RCAs ranged from seven days to eight months. While these results are without a doubt impressive, they are easily attainable when the organizational environment supports the root cause analysis (RCA) activities. Read on and become a believer.

CASE HISTORY 1:
ISPAT INLAND, INC. (EAST CHICAGO, IN)

UNDESIRABLE EVENT

Catastrophic Failure of #4 BOF (Basic Oxygen Furnace) Lance Carriage Assembly in the Steel Making Area.

UNDESIRABLE EVENT SUMMARY

During a routine slag wash, the operator in the pulpit (control room) was raising the 11-ton lance carriage. While raising the lance to its idle position approximately 80' above the fourth floor, a coupling on the drive platform failed, sending the lance carriage into a free fall. The carriage broke through the stop bolts and crashed into the 4th floor. See Figures 12.1 and 12.2.

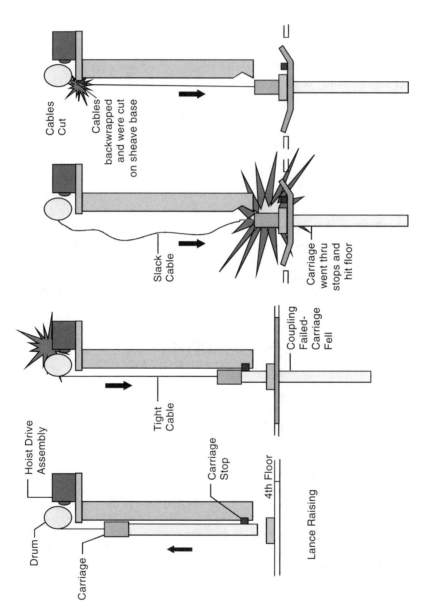

FIGURE 12.1 Lance carriage free fall.

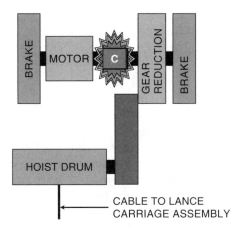

FIGURE 12.2 Hoist drive platform (top view).

LINE-ITEM FROM MODIFIED FMEA

TABLE 12.1
Line Item from Modified FMEA

Subsystem	Event	Mode	Frequency	Impact/Occurrence*	Total Annual Loss
#4 BOF	Catastrophic failure of lance carriage assembly	Lance carriage hits floor	9/2 years	$257,940	~$1,160,000

Note 1: Impacts include labor cost, material cost, and lost profit opportunities from lost sales.

IDENTIFIED ROOT CAUSES

Physical Roots

- Excessive vibration
- Insufficient motor movement prevents proper alignment
- Improper gear mesh
- Wrong hug nut installed on brake pull rod
- Debris under sheave plate could have prevented movement
- Dirt/graphite not cleaned out
- Current design allows dirt to accumulate

Human Roots

- Unacceptable conditions not observed during field inspections
- Severe misalignment
- Field inspection error

- Aware of unacceptable conditions, but not responded to
- Current safety dog system design inadequate

Latent Roots

- Torque procedures for coupling bolts nonexistent
- Inadequate alignment procedure
- Improper alignment tools available
- Lack of training in proper alignment practices
- No audit function of preventive maintenance (PM) inspections
- No audit procedure for PM inspections
- Lack of understanding entire BOF process/system
- PM checklist not on site
- Routine inspections not assigned to inspectors
- Inspections seen as undesirable job
- Correct hug nuts not available in stores
- Lack of formal training in braking systems
- New brake equipment not added to inventory
- Safety dogs activated too late
- Coupling failure not compensated for in current design
- Safety dog system designed only to activate during slack cable condition
- Sheave plate area perceived as difficult/unsafe environment to clean
- Multiple levels/floors to clean
- Excessive time and effort to set up vacuum equipment
- Cleaning perceived as downturn work
- Fewer downturns experienced over time
- PM process dictates frequency of inspections
- Perceived lack of time to train in overall BOF process
- Job conflict perceived with PM inspections
- Major repairs to other equipment was of higher priority
- Maintenance cleaning tasks perceived as low priority
- Mentality that production is to be maximized in the short term
- Perception that there was not enough time to align properly

IMPLEMENTED CORRECTIVE ACTIONS

- Precision align all components to gear reduction
- Remesh all gears
- Conduct formal training in proper alignment practices
- Institute "sign-off" of alignment on drives
- Change present bases out
- Survey fabricated bases ensuring all mounting holes are perpendicular and parallel
- Modify dry gear mesh to be enclosed with lubrication

- Change gears in sets
- Investigate "unit" exchange of drive assembly
- Conduct formal training on brake systems
- Update computerized maintenance management system (CMMS) when new equipment is installed
- Inspect and clean safety dog gap on all four lane carriages
- Improve maintenance on lance carriages
- Incorporate torque specifications for coupling bolts in alignment sign-off document
- Include torquing effects in formal alignment practices
- Rewrite alignment procedure to include special requirements and location of alignment tools and brackets
- Utilize the work order system to schedule audits of PM inspections to include standards of inspection
- Conduct brief classes on the BOF process and how the equipment functions within that process
- Provide a "checklist" carrier on site to enable timely updating of the inspections as they are being performed
- Conduct pre-job meetings in which possible job conflicts are discussed and resolved
- On critical jobs, mandate that the PM supervisor audit every time
- Reduce or eliminate the fluctuation of PM individuals assigned to the task
- Train the PM supervisor to be accountable for the quality of the job even if the individuals are on the job for the first time (again, only on critical jobs performed 100% by the PM crews)
- Develop a new/modified equipment checklist which ensures new equipment and spares are added to the CMMS (RCM tree) and subsequently, to the inventory
- Replace the overload/underload limit switch system with a system utilizing load cells
- Install new boot system on safety dog gap
- Perform process mapping analysis
- Review prioritization of FMEA/PMA master plans
- Distinguish between maintenance cleaning and housekeeping
- Challenge perceptions of downturn needs
- Complete all necessary maintenance to standards
- Investigate paradigm of "production is to be maximized in the short term"
- Conduct sessions to inform and educate everyone involved who must support the shift in mind-sets
- Hold employees accountable for deviations in product continuity, quality, safety, etc., that are manifestations of behavior stemming from the old paradigms
- Recognize those employees who demonstrate with their actions the use of the new paradigms

EFFECT ON BOTTOM LINE

Tracking Metrics

- PMs monitored weekly
- PM schedule compliance tracked weekly
- PM exceptions are investigated and countermeasures taken
- Mean time between failure (MTBF) and mean time to restore (MTTR) are tracked monthly

Bottom-Line Results

- MTBF improved from 75 days to 538 days (and counting) — 700% increase.
- Departmental PM performance is tops in the plant.
- $1,150,000 material cost reduction (1995 vs. 1997).
- Experienced a labor reduction in resources necessary to address emergency repairs. However, utilized additional labor resources to increase PM frequencies from monthly to weekly. This basically resulted in the moving of reactive work to proactive work.

Corrective Action Time Frames

- Most management systems (latent roots) were addressed immediately.
- The physical redesign and installation of the upgraded lance carriage system took approximately 18 months.
- The final countermeasures of all recommendations were completed on June 30, 1998.

RCA Team Statistics

Start Date: November 16, 1996
End Date: June 26, 1997
Estimated Cost to Conduct RCA: $30,000
Estimated Returns from RCA: $1,150,000
Return on Investment: ~4000%

RCA TEAM ACKNOWLEDGMENTS

Principal Analyst: John Van Auken
Title: Day Supervisor-Maintenance
Company: ISPAT Inland, Inc.
Department: #4 BOF/#1 Slab Caster/RHOB
Site: East Chicago, IL
Core RCA Team Members:
  Jeff Jones
  Jim Modrowski
  Mike Sliwa

ADDITIONAL RCA TEAM COMMENTS

The training and support from RCI during the RCA and their doggedness helped make this effort work. The real benefactors are ISPAT Inland, Inc. and their maintenance organization. We go about our business differently as a result of this RCA experience. It is not "assumed" that is the way it happened any more. We "deep drill" and come up with better countermeasures. #4 BOF is recognized as a leader in RCA and maintenance methodology at ISPAT Inland. Other departments are calling us for our ideas and advice. We are proud of this accomplishment. We cannot and will not rest on our laurels. There are other opportunities here and my hope is to see ISPAT Inland use this methodology even more and permanently eliminate more of our problems utilizing the 80/20 rule. RCI's RCA methods are foolproof and proven. The proof is this RCA and its results.

John Van Auken
ISPAT Inland
RCA Prinicpal Analyst

LOGIC TREE

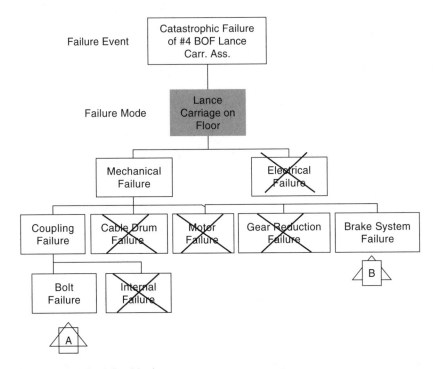

FIGURE 12.3 Inland Steel logic tree.

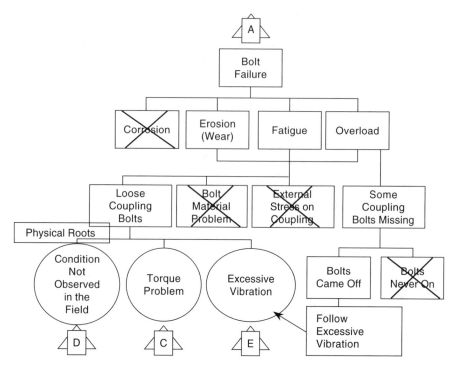

FIGURE 12.3 (continued) Inland Steel logic tree.

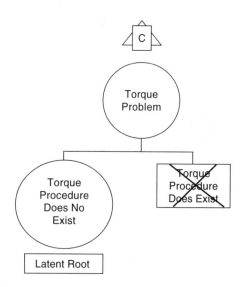

FIGURE 12.3 (continued) Inland Steel logic tree.

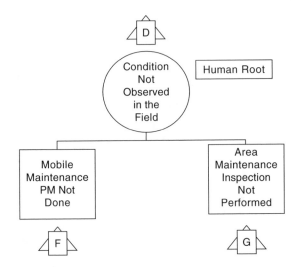

FIGURE 12.3 (continued) Inland Steel logic tree.

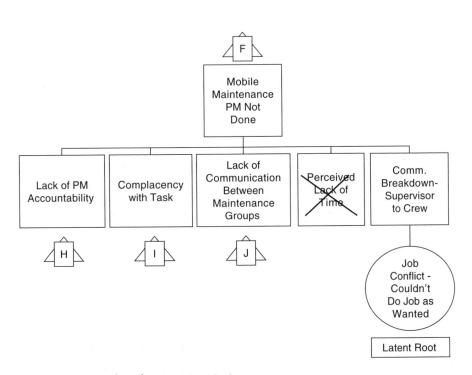

FIGURE 12.3 (continued) Inland Steel logic tree.

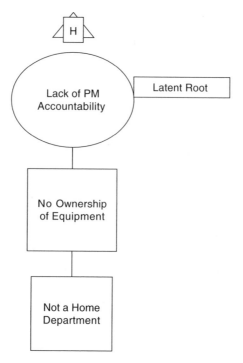

FIGURE 12.3 (continued) Inland Steel logic tree.

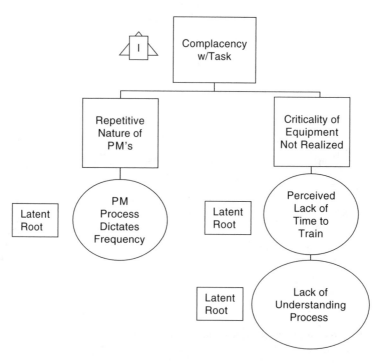

FIGURE 12.3 (continued) Inland Steel logic tree.

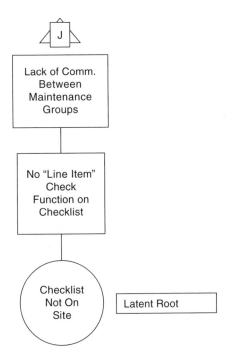

FIGURE 12.3 (continued) Inland Steel logic tree.

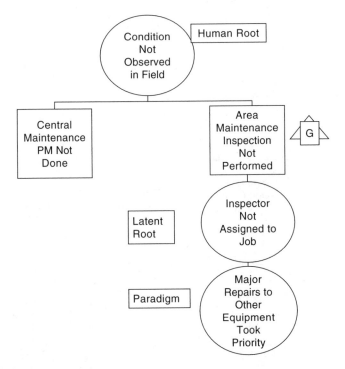

FIGURE 12.3 (continued) Inland Steel logic tree.

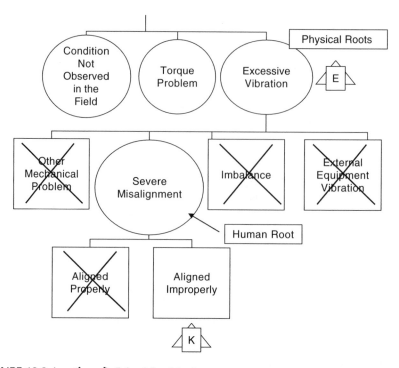

FIGURE 12.3 (continued) Inland Steel logic tree.

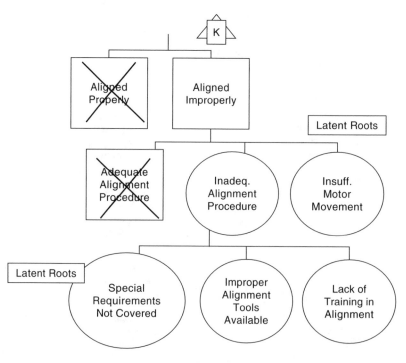

FIGURE 12.3 (continued) Inland Steel logic tree.

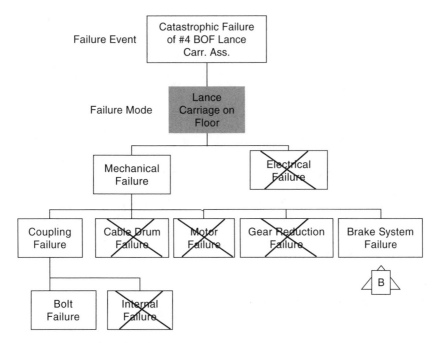

FIGURE 12.3 (continued) Inland Steel logic tree.

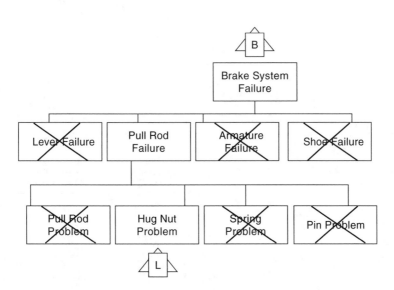

FIGURE 12.3 (continued) Inland Steel logic tree.

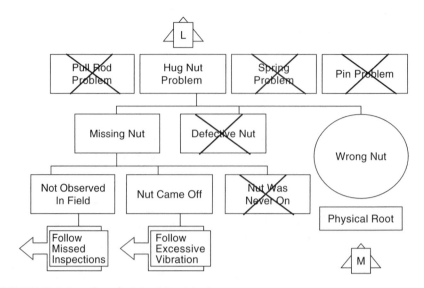

FIGURE 12.3 (continued) Inland Steel logic tree.

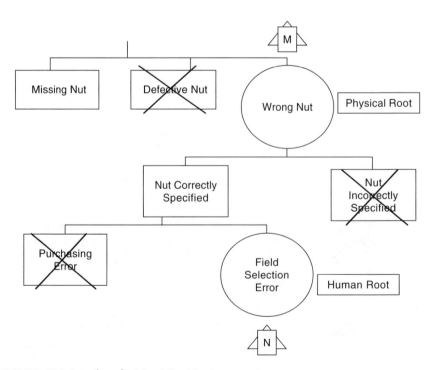

FIGURE 12.3 (continued) Inland Steel logic tree.

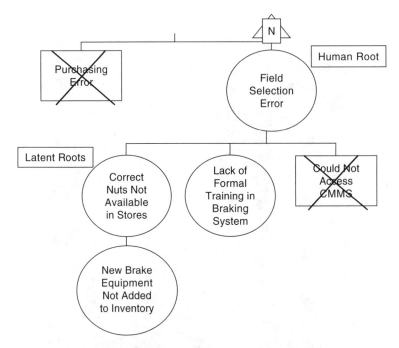

FIGURE 12.3 (continued) Inland Steel logic tree.

CASE HISTORY 2:
EASTMAN CHEMICAL COMPANY (KINGSPORT, TN)

UNDESIRABLE EVENT

Customer Complaints

UNDESIRABLE EVENT SUMMARY

Five similar customer complaints were received concerning green pellets mixed with clear pellets. Complaints were received from more than one customer, but not all rail cars of product received a complaint.

The silos and conveying systems were checked prior to their initial use for the clear product. They were also cleaned and inspected after each customer complaint. Each time, one or more potential sources of green contamination was found and corrected.

After the fifth complaint, a team was put together to discover and eliminate the root cause of the contamination. See Figure 12.4.

LINE ITEM FROM MODIFIED FMEA

See Table 12.2 for an example of a line item from a modified FMIA.

FIGURE 12.4 Product silo with blend tubes.

TABLE 12.2
Line Item from Modified FMEA

Subsystem	Event	Mode	Frequency	Impact/ Occurrence*	Total Annual Loss
Customer service	Customer complaints	Green pellets mixed with clear pellets	5 railcars in 7 months (190,000 #/railcar)	$17,100	$85,500

* *Note 1*: Impacts include labor cost, material cost, and lost profit opportunities from lost sales.

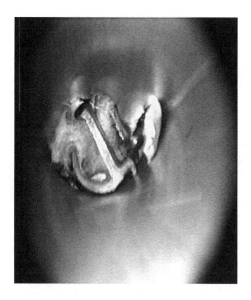 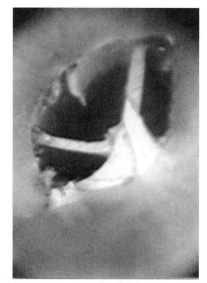

FIGURE 12.5 Plugged blend tubes.

IDENTIFIED ROOT CAUSES

Physical

- One of the silo blend tubes was damaged, causing green pellets to be held in place and released intermittently. See Figure 12.5.

Human Roots

- Poor repair process was used to patch the broken blend tube
- Inadequate cleaning
- Inadequate inspection of silo

Latent Roots

- Blend tube/support design allowed fatigue failure
- Cleaning and inspection process inadequate and poorly documented

IMPLEMENTED CORRECTIVE ACTIONS

- Damaged blend tube thoroughly cleaned
- Cleaning/inspection procedures developed and documented
- Blend tube repair procedure developed and documented
- Communicate new procedures to operations and maintenance personnel

- Conveying system/silo product changeover check sheet developed and deployed
- An improved blend tube design used in new silos

EFFECT ON BOTTOM LINE

Tracking Metrics

- Number of customer complaints concerning green pellets

Bottom-Line Results

- Zero customer complaints since root cause was found and countermeasures implemented.
- Conservative estimates report the damaged blend tube held enough green pellets to contaminate five more railcars of clear product.

$$(5 \text{ Railcars}) \times (190,000 \text{ lbs./Railcar}) \times (\$0.09/\text{lb.}) = \$85,500$$

Corrective Action Time Frames

- From first complaint to correction was seven months
- RCA team found and corrected root causes in seven days

RCA Team Statistics

Start Date: July 14, 1998
End Date: July 21, 1998
Estimated Cost to Conduct RCA: $2,700
Estimated Returns from RCA: $85,500
Return on Investment: ~3200%

RCA TEAM ACKNOWLEDGMENTS

Principal Analyst: Kevin Bellamy
Title: Reliability Engineer
Company: Eastman Chemical Company
Department: Reliability Technology
Site: Kingsport, TN
Core RCA Team Members:
  Leslie White
  Keith Bennett
  Lee Norell
  Michael Lambert

Logic Tree

See Figure 12.6.

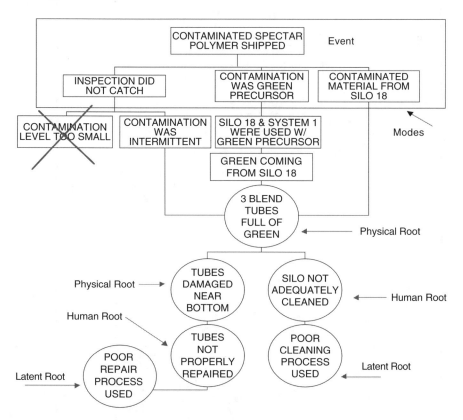

FIGURE 12.6 Eastman Chemical logic tree.

CASE HISTORY #3:
LYONDELL-CITGO REFINING (HOUSTON, TX)

Undesirable Event

Vacuum Column Bottoms Pump Failure

Undesirable Event Summary

Recurrent failures of vacuum column bottom pumps. Both pumps came on line in December 1996. The mean time between failure (MTBF) was very poor at three months. Failures of mechanical seals, thrust bearings, impellers, and case wear rings were very common. See Figure 12.7.

Most of the failures occurred at start-up. The system operates with one pump as a primary pump and the other as a spare pump. Different attempts to correct the

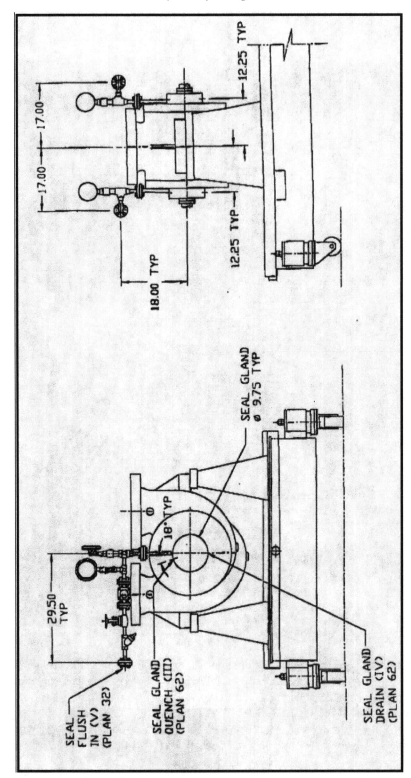

FIGURE 12.7 Vacuum column bottom pump.

above problems failed. There was not a good understanding of the causes of these failures or most important, how they correlated to each other. At times, both pumps would be unavailable. The impact on production and the excessive maintenance costs resulted in management appointing a root cause analysis (RCA) team to find and implement final solutions to these problems.

LINE ITEM FROM MODIFIED FMEA

See Table 12.3.

TABLE 12.3
Line Item from Modified FMEA

Subsystem	Event	Mode	Frequency	Impact/ Occurrence*	Total Annual Loss
Vacuum column	Bottoms pump failures	Seal failure, bearing failure, and wear ring failure	5/Yr	$1,431,000	~$7,150,000

* *Note 1:* Impacts include labor cost, material cost, and lost profit opportunities from lost sales.

IDENTIFIED ROOT CAUSES

Physical Roots

- Cooling water line plugged
- Suction and discharge pipes plugged
- Steam trap not working
- Inadequate clearance
- Uneven thermal growth
- Loose wear rings
- Minimum flow line blocked in
- Heat checking on seal faces
- Steam tracing not working

Human Roots

- Inadequate design: warm-up lines too small/not enough heat tracing
- Inadequate warm-up: no temperature check before start-up
- Improper start-up: cold start-up, pump operating at dead-end for a long time
- Improper installation
- Steam trap blocked in

Latent Roots

- Inadequate warm-up systems
- Incorrect specifications and procedures
- Inadequate training on bearing installation
- Lack of start-up/shutdown procedures
- Inadequate operating procedures/training

IMPLEMENTED CORRECTIVE ACTIONS

- Install electrical tracing on suction/discharge pipes
- Revise cooling water line from series to parallel
- Enlarge warm-up line from ½" to 2" diameter
- Revise seal flush
- Relocate flush line from a 500'F source to less than 200'F source
- Revise standard operating procedure (SOP) and train operators on new start-up/shutdown procedures

EFFECT ON BOTTOM LINE

Tracking Metrics

- Mean time between failures (MTBF) increased from 3 months to 11 months

Bottom-Line Results

- New start-up/shutdown procedures have proved to be successful
- Large warm-up lines have avoided blockage
- Eliminated impeller wear rings so those failures have been eliminated
- Replaced seal flush with a cooler source
- Pump warm-up is controlled by electric tracing with digital read-out for a total of 14 check points
- No more cooling water line blockage
- Estimated savings of $7,150,000 ($6,500,000 in production losses and maintenance labor and material costs of $655,000)

CORRECTIVE ACTION TIME FRAMES

- Total of five months
- The RCA team expended two months
- The recommendation implementation took three months

RCA Team Statistics

Start Date: August 4, 1997
End Date: September 26, 1997
Estimated Cost to Conduct RCA: $40,000
Estimated Returns from RCA: $7,150,000
Return on Investment: ~17,900%

RCA Team Acknowledgments

RCA Sponsor: Jimmy McBride
Title: Manager, Mechanical Support and Reliability
Company: Lyondell-Citgo Refining
Department: Reliability Engineering
Site: Houston, TX

Principal Analyst: Edgar Ablan
Title: Principal Engineer
Company: Lyondell-Citgo Refining
Department: Reliability Engineering
Site: Houston, TX

Core RCA Team Members:
 Terry Dankert
 David Collins
 Mahesh Patel

Additional Comments

The effort of this cross-functional team using the RCI method has proven that focusing on implementing solutions to the root causes of failures will improve equipment reliability and generate very attractive savings.

Jimmy McBride
LYONDELL-CITGO Refining
Houston, TX

Logic Tree

See Figure 12.8.

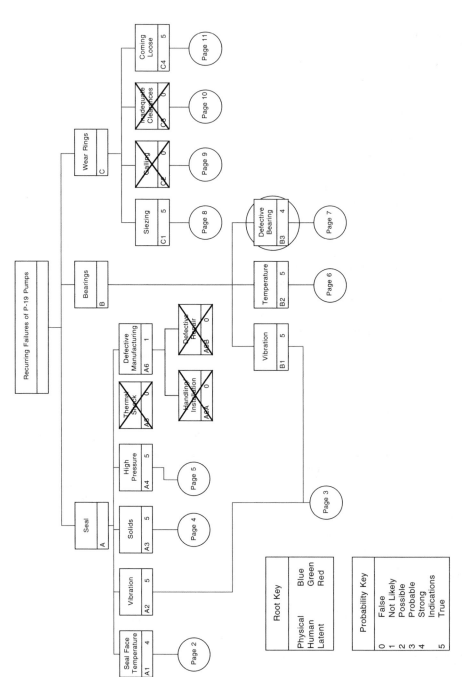

FIGURE 12.8 LYONDELL-CITGO logic tree.

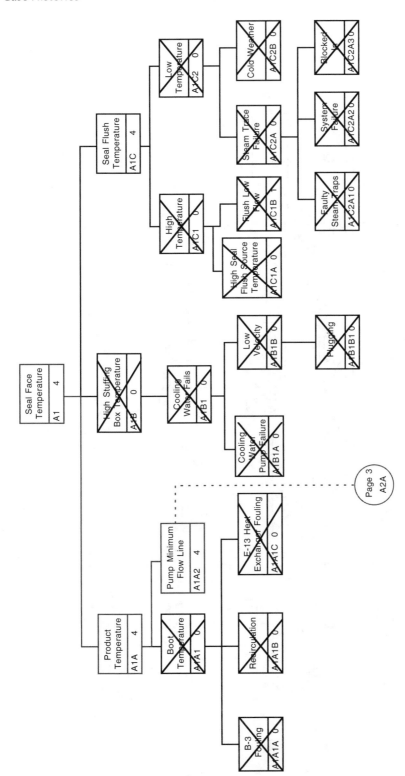

FIGURE 12.8 (continued) LYONDELL-CITGO logic tree.

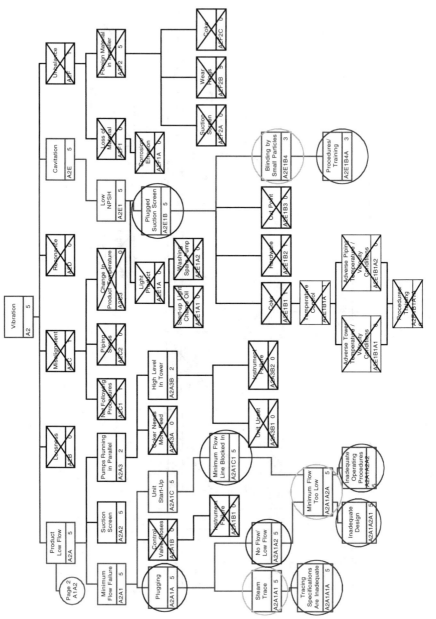

FIGURE 12.8 (continued) LYONDELL-CITGO logic tree.

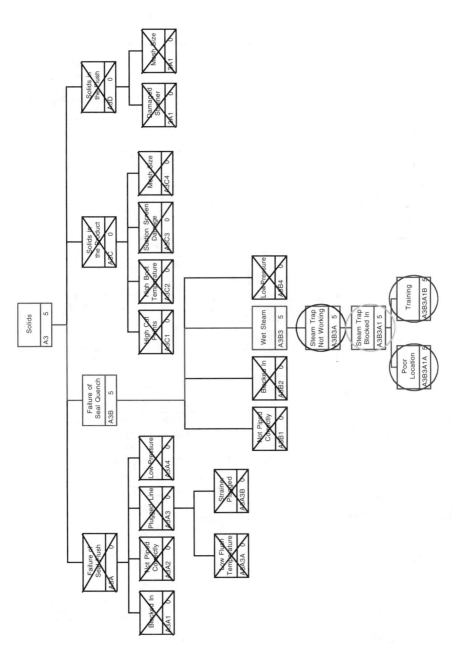

FIGURE 12.8 (continued) LYONDELL-CITGO logic tree.

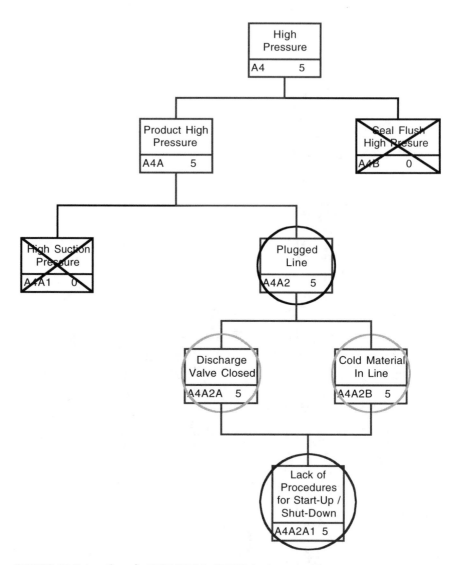

FIGURE 12.8 (continued) LYONDELL-CITGO logic tree.

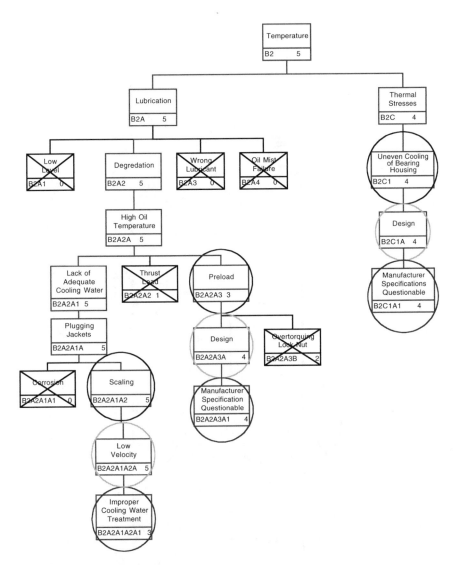

FIGURE 12.8 (continued) LYONDELL-CITGO logic tree.

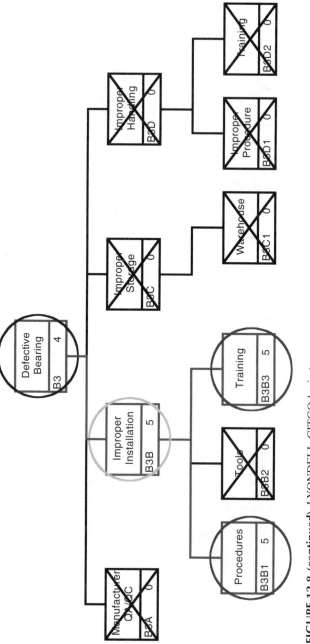

FIGURE 12.8 (continued) LYONDELL–CITGO logic tree.

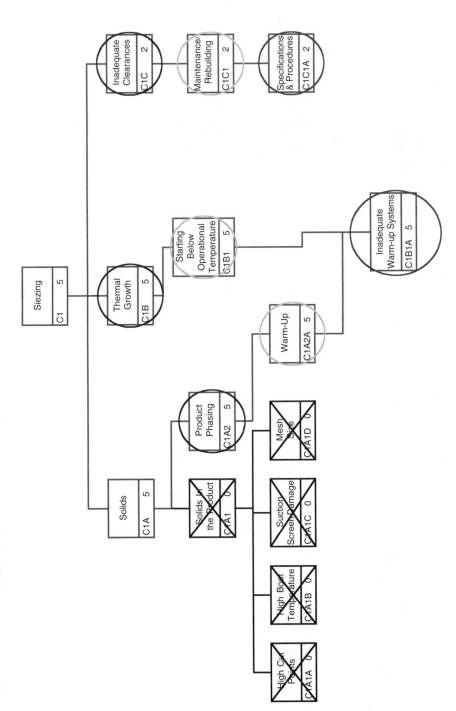

FIGURE 12.8 (continued) LYONDELL-CITGO logic tree.

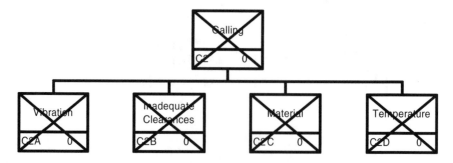

FIGURE 12.8 (continued) LYONDELL-CITGO logic tree.

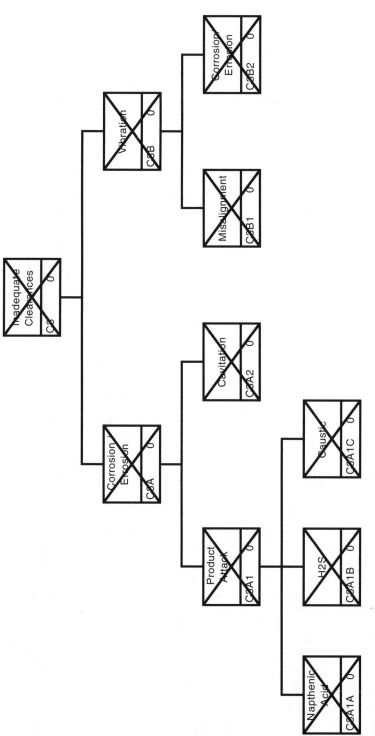

FIGURE 12.8 (continued) LYONDELL-CITGO logic tree.

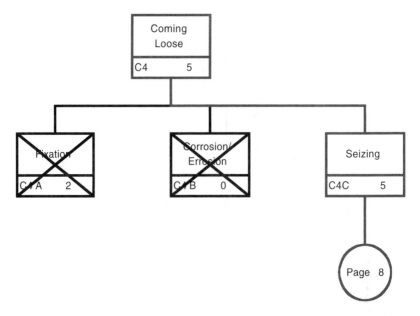

FIGURE 12.8 (continued) LYONDELL-CITGO logic tree.

CASE HISTORY 4:
EASTMAN CHEMICAL COMPANY
(WORLD HEADQUARTERS: KINGSPORT, TN)

UNDESIRABLE EVENT

Unacceptable Level of Reduction of Worldwide Customer Complaints

UNDESIRABLE EVENT SUMMARY

During 1997, Eastman Chemical senior management realized that their level of customer complaints had not shown significant reduction during the past few years. This was troubling, given that Eastman had a strong history of continually improving performance in their processes. Furthermore, one of the key objectives of Eastman's Customer complaint handling process was to investigate and identify the cause of complaints — evidently the complaint investigations were not as effective as expected.

A team was appointed which then studied the complaint investigations that had been occurring at Eastman, and the team discovered that most complaint investigations were not getting to the root (organizational) causes. Rather, most investigations had stopped at who caused the problem. Consequently, the corrective action plans were typically along the lines of we will pay more attention in the future, we will be more careful, we will try harder, etc.

Eastman recognized a more thorough identification of the root causes of complaints was needed. Eastman's customer complaint advocates and complaint investigators at their sites worldwide needed to better understand the appropriate methodology to more thoroughly identify causes of complaints, so that the appropriate actions could be taken to eliminate the causes of recurring complaints.

Eastman turned to Reliability Center Incorporated (RCI) to help them develop an RCA training course for Eastman people worldwide to more thoroughly understand how to identify and therefore eliminate causes of complaints, especially recurring complaints. Eastman's customer complaint manager received the "train-the-trainer" training in Hopewell in February 1998. During the remainder of 1998, he then provided this training to over 300 people located at every Eastman site worldwide.

Additionally, complaint reduction through defect prevention was made a corporate initiative. This involved much management support throughout the entire organization, which of course provided needed focus to the effort.

Much measurement supported this corporate initiative, where progress would be measured in terms of number of complaints per million shipments (PPM — parts per million shipments). Each organization adopted this measurement. The goal was established to reduce the level of complaints to half the 1997 level, through defect prevention, over a three-year time frame. Much monitoring occurred, and much positive reinforcement was provided where appropriate.

LINE ITEM FROM MODIFIED FMEA

While conventional measurement information for the modified FMEA was not available in this case, we can imagine the bottom-line effect that reducing worldwide complaints by 50% would have.

The cost of complaints is very significant, manifested in:

- Lost business when customers switch to other suppliers
- Handling costs associated with responding to (including investigating) complaints
- Claims paid and credits given to customers to compensate customers for added costs associated with their complaint
- Waste and rework associated with producing off-quality product, correcting paperwork errors, etc.

Although it is difficult to precisely calculate all monetary savings resulting from efforts involving RCA, it is estimated that bottom-line benefits to Eastman Chemical Company since 1997 are in the range of several million dollars.

SPECIFIC RCA DESCRIPTION

A customer complaint was entered due to a tank truck shipment of N-butyl alcohol being transferred by the delivery agent into the customer's incorrect tank, which contained ethyl acetate. Figure 12.9 is the logic tree, conducted by the delivery

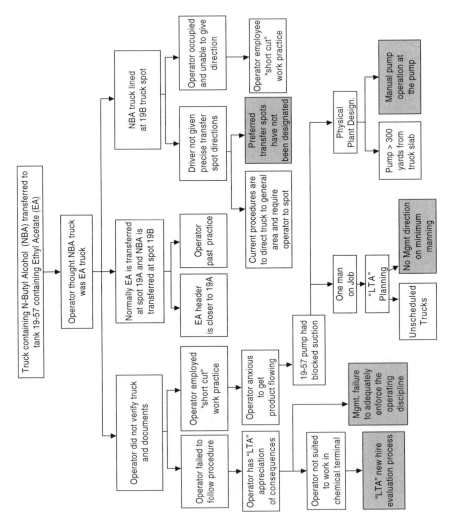

FIGURE 12.9 BayTank logic tree.

agent*. The agent found the human cause as the operator failed to follow procedure. The agent recognized, via the recently completed RCFA training, that he needed to dig deeper to discover the organizational root causes. So he kept developing hypotheses until the real organizational root causes were discovered for this — (1) a less-than-acceptable process for evaluating new hires and (2) management failure to adequately enforce the operating discipline. Other organizational causes were also found for other legs. Actions were implemented, and performance has since been very favorable.

IDENTIFIED ROOT CAUSES

Major Physical Roots

- Physical transfer N-butyl alcohol into ethyl acetate tank

Major Human Roots

- Operator did not verify truck and documents
- Driver not given precise transfer spot directions
- Operator occupied and unable to give direction

Major Latent Roots

- Less than acceptable (LTA) new hire evaluation process
- Management failure to adequately enforce the operating discipline
- No management direction on minimum manning
- Preferred transfer spots have not been designated
- Manual pump operation at the pump

IMPLEMENTED CORRECTIVE ACTIONS

- Developed and implemented new hire evaluation process
- Operating discipline enforced due to management commitment
- Minimum manning requirements set and communicated
- Preferred transfer spots designated

EFFECT ON COMPANY BOTTOM LINE

Tracking Metrics

- Complaint progress is tracked monthly via the measurement "number of complaints per million shipments"

---

* Baytank (Houston), Inc. 12211 Port Road, Seabrook, TX, 77586.

Bottom-Line Results

- Elimination of half of Eastman's customer complaints. This equates to about $2,000,000 on reduced complaint handling costs, reduced waste and rework, and not losing customers due to poor quality.
- Reduced operating and maintenance costs, from improved process and equipment reliability
- Reduced account receivables
- Improved organizational effectiveness

Corrective Action Time Frames

- Continuous improvements from 1998 to 2001 (see Figure 12.10)

RCA ACKNOWLEDGMENTS

Principal Analyst: Gary Hallen
Title: Customer Focus Manager
Company: Eastman Chemical Company
Scope: Global Customer Service Group
Acknowledgments: We would like to thank Baytank, Inc. for use of their RCA example (Figure 12.9). We admire their drive to conduct the RCA and their courage to let others learn from their success.

Analyst: Sam Dufilho
E-mail: sam.dufilho@baytank.com
Phone: 713-844-2300

CASE HISTORY 5:
SOUTHERN COMPANIES ALABAMA POWER COMPANY
(PARRISH, AL)

UNDESIRABLE EVENT

Recurring Failure of Unit 10 Electric Fire Pump

UNDESIRABLE EVENT SUMMARY

The Unit 10 Electric Fire Pump had failed five times within a six-month period. All of these events were due to an outboard bearing failure.

LINE ITEM FROM MODIFIED FMEA

See Table 12.4.

Event: Pump failure
Mode: Bearing failure

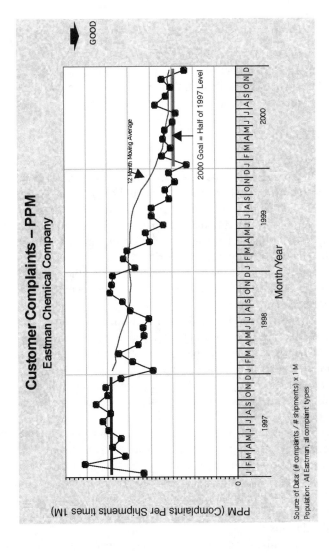

FIGURE 12.10 Eastman worldwide customer complaint reductions.

TABLE 12.4
Line Item from Modified FMEA

Event	Mode	Frequency/Yr	Impact Overall (Sum of Work Orders)	Total Annual Loss
Recurring failure of #10 electric fire pump	Outboard bearing failure	5	$25,816	$25,816

Frequency: Estimated number of occurrences/year
Impact: Sum of all work orders issued to repair this pump this year
Total Annual Loss: In this case this number is the same as the impact because all the work orders were summed vs. separated per event.

SPECIFIC RCA DESCRIPTION

After a team review of how the PROACT® process worked, a logic tree was developed to graphically represent the cause and effect relationships that could have led to the event occurring. All of the hypotheses that could be eliminated based on hard facts were ruled out and some team assignments were made for those hypotheses that still required verification. During the team meetings it was brought out that the outboard bearing was the thrust bearing. It was also explained to the other team members how they could identify a roller bearing that was a thrust bearing. Team members were assigned to retrieve the failed bearings and start the process of finding out why it failed. Team members were also assigned to locate an instruction manual and any other documentation possible.

It turned out that the old bearing was not a thrust bearing after all. The mechanics had been replacing the burned up bearings with the same type they had removed from the pump, which was the wrong bearing. Further investigation revealed that the CMMS (computerized maintenance management system) listed only one bearing for this fire pump and its description did not indicate it was the inboard bearing. The storeroom did have the correct bearing for the outboard but it was not listed for this pump.

The new bearing was installed and the pump was returned to service but was immediately removed because of high vibration. The team was called back together and continued the process to find the cause of this high vibration. A pump expert with Southern Company was on the plant site and agreed to meet with the RCA team. After the team listed the things that could cause this high vibration, the pump expert did an excellent job of explaining the proper process for building this pump and things we should look for.

The CBM (condition-based maintenance) team leader joined the RCA team meeting and gave a good account of the type of vibration found. The team identified shaft run out as a possibility because the shaft had been welded on when a new sleeve had been installed. The packing gland fit had not been as good as it should have been so it was added as a possibility. It was also decided that the impeller run out should be checked and it was noted that the impeller was not perfectly centered in the case. The coupling was thought to have been set up without the motor being at magnetic center. If this were to be accomplished, a spacer would have to be installed. A team member identified this as the first thing to check. This was found to be one of the root causes. A spacer was installed, the coupling was set up properly and the pump has run with excellent results ever since.

IDENTIFIED ROOT CAUSES

Major Physical Roots

- Wrong bearing installed for the current service
- Coupling not sliding

Major Human Roots

- Wrong set-up in CMMS for bearings
- Coupling not set up correctly

Major Latent Roots

- No parts list in CMMS
- No procedure for setting up coupling properly

IMPLEMENTED CORRECTIVE ACTIONS

- Changed the description for the inboard bearing in the CMMS to read inboard bearing
- Set up the outboard bearing in CMMS with an improved description
- Purchased an accurate instruction manual for the fire pump
- Added a parts list to CMMS for the fire pump
- Added a procedure to CMMS with an accurate and user-friendly checklist for the fire pump
- Trained mechanics and E&I on proper alignment and procedures for checking magnetic center

EFFECT ON COMPANY BOTTOM LINE

Tracking Metric

- Frequency of occurrence
- Decreased maintenance costs

Bottom-Line Results

- Elimination of the recurrence of the outboard bearing failures on #10 Electric Fire Pump since the recommendations were implemented

Corrective Action Time Frames

- The RCA was completed on January 30, 2001, and several of the recommendations were implemented before starting the pump up again. All of the recommendations were implemented by March 15, 2001. The pump has not failed since.

RCA Team Statistics

Start Date: January 22, 2001
End Date: January 30, 2001
Estimated Cost to Conduct RCA: $800
Estimated Returns from RCA: $25,816
Return on Investment: ~3125%

RCA Acknowledgments

Principal Analyst: Ronny Johnston
Title: Maintenance Planner
Company: Southern Companies
Division: Alabama Power Company
Site: Parrish, AL
Core RCA Team Members:
    Paul Cooner
    Chris Curow
    Harold Dobbins
    Steve Newton
    David Hosmer
    MC2 Maintenance Team
    Warehouse Team

Logic Tree

See Figure 12.11.

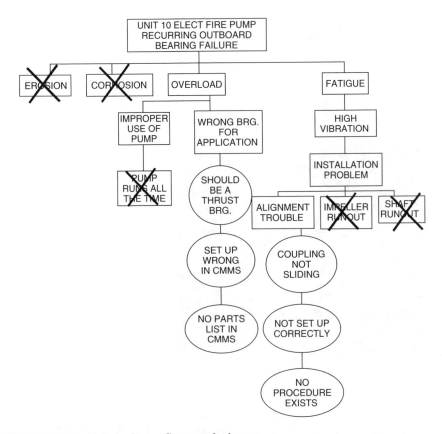

FIGURE 12.11 Alabama Power Company logic tree.

CASE HISTORY 6:
WEYERHAEUSER COMPANY (VALLIANT, OK)

UNDERSIRABLE EVENT

Catastrophic Failure of the Thermo Compressor Cone for the Number 3 Paper Machine

UNDESIRABLE EVENT SUMMARY

New equipment for Paper Machine Number 3 was started up on March 1, 2000. After about two months service a crack developed around the gauge port of the thermo compressor propagating longitudinally from the toe of the fillet weld, around the gauge port, into the base material of the thermo compressor (Figure 12.12).

The thermo compressor was replaced with a like component. After two days' service, a leak developed at the longitudinal weld seam similar to the initial failure of the original thermo compressor (Figure 12.13).

Thermo compressor piping was visually inspected and subsequently analyzed because of the repeated failures of the cone attachments and the component body.

Crack Origin

Note that the crack branches turn in the lengthwise direction

FIGURE 12.12 Crack origin.

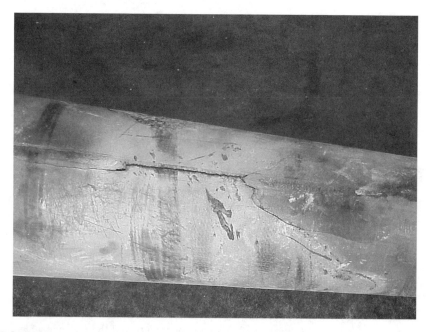

FIGURE 12.13 Crack along longitudinal seam of thermo compressor cone.

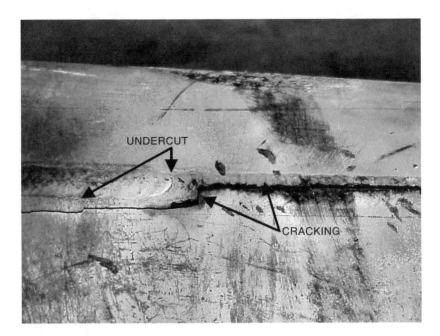

FIGURE 12.14 Weld defects in thermo compressor.

In doing so, a conventional piping analysis of the design considering gravity, pressure, and thermal growth forces showed that they were well within the appropriate standards during normal operation, both with and without the thermo compressor in service. Flaws in the pipe elbow of the thermo compressor were shown to progress into the piping wall and were, therefore, quite serious. In addition, visual inspection of the weld area revealed defects that contributed to the failure, including cracking and undercut (Figure 12.14).

Micro and macro metallurgical examination of the weld area and heat-affected zone uncovered additional weld application and heat-treating defects that accounted for the stepping appearance of the cracks. These examinations showed that the cracking followed a pattern that stepped from filler and base metal inclusions (Figure 12.15) on a background tempered martensite base metal microstructure. In essence, the crack approximated a pattern that represented flaws in the cone's metallurgical microstructure, or the path of least resistance throughout the filler and base metals of the thermo compressor's cone-shaped body.

Visual examination of the weld's crack cross-sectional area showed evidence of fatigue contributing to the failure of the thermo compressor cone. In Figure 12.16, the arrow shows the direction of the crack in the base metal. In addition, upon further close examination you can see circular beach marks that fan out from the outside diameter of the cone. This combined with the ratchet marks and a generally flat surface are clear indications of a fatigue-related failure demanding further examination of the process.

To confirm fatigue as a contributor to the failure being analyzed, pipe wall flexure natural frequencies in the piping that constitute the thermo compressor were identified

FIGURE 12.15 Microstructure defects.

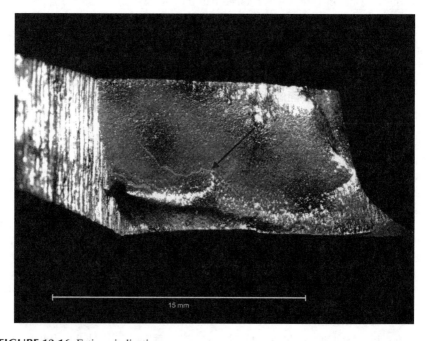

FIGURE 12.16 Fatigue indications.

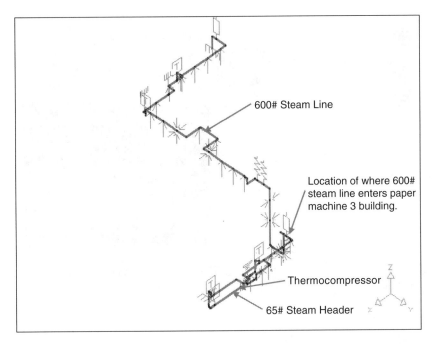

600# Steam Line

Location of where 600#
steam line enters paper
machine 3 building.

Thermocompressor

65# Steam Header

FIGURE 12.17 Piping system with 65 psig steam header excluding expansion loop.

with numerical methods and confirmed through testing by plant personnel in the field under operating conditions. Here it was determined that the natural frequency mode shapes were consistent with the location and orientation of the cracks. Furthermore, all failure modes — attached cracking, loosening nuts, and cracked cone — are consistent with the vibration induced from thermo compressor operation. The analytical conclusion of the source of the fluctuating stress that was producing the fatigue failures was determined to be pipe wall resonance, and that any coincidental acoustical resonance would synergistically magnify this vibratory stress.

The system analyzed for gravity, pressure, and thermal stress is illustrated in Figure 12.17. The existing 65 psig steam header was limited in the model because it appeared to be unnecessary.

Piping stress in and around the thermo compressor was determined to be low, and there was seemingly no correlation between the failures and piping stress from gravity, pressure, and thermal loadings.

Vibration of the piping from which the thermo compressor cone is constructed was analyzed by constructing a finite element computer model. Several natural frequencies were identified by dynamic analysis. The natural frequency mode shape that occurs at 540 Hz (seen in Figure 12.18) shows a pipe wall flexure that will produce the highest stress at precisely the location of an experienced crack. In addition, the stress will also fluctuate, which is a necessary prerequisite for fatigue cracking.

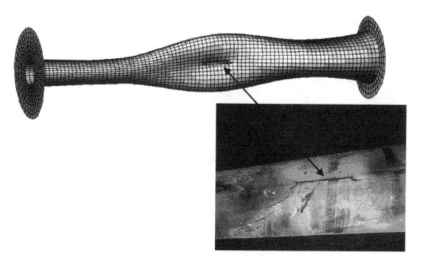

FIGURE 12.18 Dynamic finite element model of the thermo compressor cone showing a fundamental 540 Hz natural frequency pipe wall flexure mode

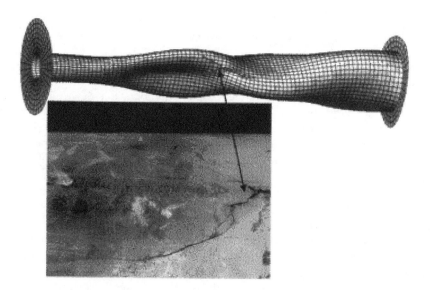

FIGURE 12.19 Natural frequency mode at 777 Hz.

The mode shape of Figure 12.18 does not explain the initial crack branching into two cracks at the left end. The mode shape of Figure 12.19 shows the initial crack running into high fluctuating stress fields that are at nominally 45 degrees on either side. The concave and convex areas alternate and provide the fluctuating high stress fields necessary for the fatigue crack to advance in their directions.

Field measurement of the thicker cone showed a strong, undamped resonance at 780 Hz (46,000 cpm). This coincides with the 777 Hz natural frequency of the pipe wall and explains the axial crack branching to higher stress fields.

The piping elbow that forms the thermo compressor had weld flaws that extended into the wall of the elbow (Figure 12.20).

Spring hangers were not properly adjusted because the basis for adjusting the support system is unclear. The design calls for a cold setting, which is defined as total shutdown of the system, and a hot setting that is obviously with the thermo compressor in operation.

It was observed that the mechanical fasteners for the thermo compressor flange near the spectacle blind tie-in at the 65 psig header were loosening during operation due to high frequency piping vibrations. The piping vibration and high Db noise levels from the thermo compressor are proportionately amplified by excitation of the thermo compressor's structural natural frequencies, especially in concert with acoustical natural frequencies. This contributed to the creation of a steam leak due to gasket or flange facing damage from previous operation with loose mechanical fasteners.

In general, static loads are acceptable by engineering code. Failure cannot be contributed to the static loads induced by the system, but by a fatigue mechanism. In addition, cracking failures on the thermo compressor cone are related to pipe wall flexure resonance that is excited by normal thermo compressor noise and vibration. Any coincidental acoustical natural frequencies, or their harmonics, will accentuate vibratory stress.

The quality issues addressed earlier are significant for improving the life of the thermo compressor. The margin of safety on the thermo compressor cone is unknown at this moment but can be determined with an engineering assessment involving quantitative dynamic finite element analysis for stress with fatigue considerations.

LINE ITEM FROM MODIFIED FMEA

IDENTIFIED ROOT CAUSES

Physical Roots

- Inadequate supports for thermo compressor and associated piping
- Defective thermo compressor base metal
- Shop welds defective
- Condensate in the system because of control valve positioning on the 3rd and 4th sections
- Incomplete fusion
- Condensate drains impeded by back pressure in the system

Human Roots

- Support system design error
- Thermo compressor not specified correctly
- Weld application error

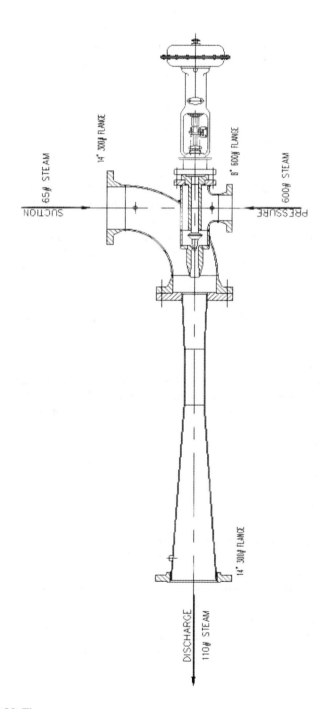

FIGURE 12.20 Thermo compressor.

- Design deficiency of condensate system
- Weld technique defective
- Running at low/high turndown ratios

Latent Roots

- Inadequate component specifications for support system
- Vendor did not understand system operating environment
- Inadequate specifications supplied to vendor for thermo compressor
- Original design of condensate traps inadequate for service
- No weld procedure specification
- Varying operating speeds to meet customer/plant requirements
- Did not follow weld procedure
- No heat treatment requirements for thermo compressor

IMPLEMENTED CORRECTIVE ACTIONS

- New specifications for permanent replacement thermo compressor cone to include:
  - Base metal to be ½ inch chrome moly material instead of 5/16 inch grade 516 carbon steel
  - 100% radiographic examination (x-ray) for all welds
  - Delete installation of gauge port from the thermo compressor
  - Stress relieve the assembly after fabrication
- Require thermo compressor manufacturer to supply:
  - Welding procedure specification used in manufacturing
  - Radiographic film showing all weld passes per ASME B31.1 & ASME Section IX
  - Stress-relieving procedure per ASME section IX
- Conduct an engineering assessment to determine the margin of safety against thermo compressor cone failure — methodology and calculations should be well documented and open to critical review.
- Take vibration measurements for amplitude and frequency to analyze piping both before and after thermo compressor startup.
- Inspect for steam leaks at the thermo compressor flanges during startup and periodically during operation.
- Modify piping and associated supports for the thermo compressor in a suspended manner with the required clearance between the piping and the support structure as indicated by the outcome of the stress analysis.
- Adjust spring hangers and note and mark the hangers to reflect both cold and hot positions.
- Install a thermal well in the 600# steam to the thermo compressor to monitor the stability of the steam temperature at the point of use.
- Reroute the drainage of the condensate traps from the 65#, 600#, and 120# steam piping to minimize the effects of backpressure and steam/water hammer in the condensate drainage system.

- Revise operating procedures to limit the thermo compressor turndown ratio between the 65# inlet and outlet to be less than 1.80 to mitigate continuous surging when reaching the theoretical limit of the thermo compressor.

Effect on Bottom Line

Tracking Metrics

Production capacity increase

Bottom-Line Results

25% Increase in production capacity

Corrective Action Time Frames

Approximately 4 months

RCA Team Statistics

Start Date: May 1, 2000
End Date: May 8, 2000
Estimated Cost to Conduct RCA: $41,476
Return on Investment: 1040%

RCA Team Acknowledgments

Principal Analyst: Ronald L. Hughes
Title: Senior Reliability Consultant
Company: Reliability Center, Inc.
Core RCA Team Members:
  Douglas Dretzke, Weyerhaeuser
  Matt Connolly, Weyerhaeuser
  Steven Breaux, Weyerhaeuser
  Theron Henry, Weyerhaeuser
  Joel White, Weyerhaeuser
  Freddy Rodriguez, Kellogg Brown and Root

Additional RCA Team Comments

"Ron Hughes has done a tremendous job for Weyerhaeuser at Valliant and represented RCI in a highly professional manner. I believe that his contributions have advanced our skills and will enhance our future profitability. We are looking forward to an ongoing relationship to all of you at RCI."

Joel White
Weyerhaeuser
Preventive Maintenance Engineer

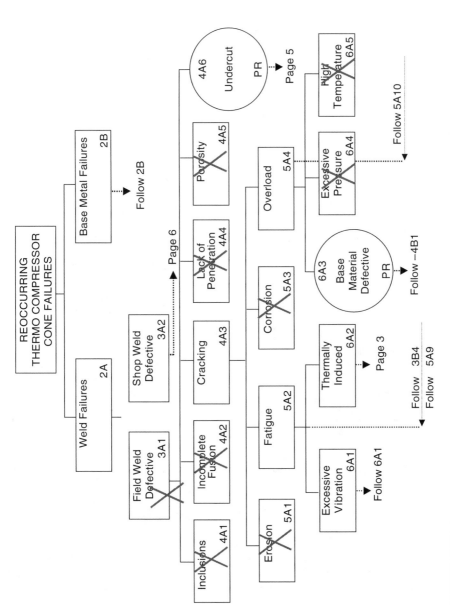

FIGURE 12.21

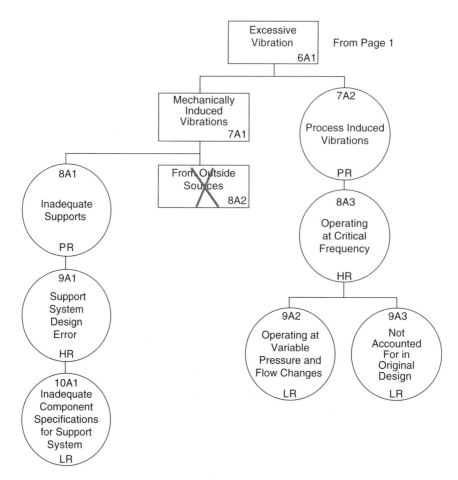

FIGURE 12.21 (continued)

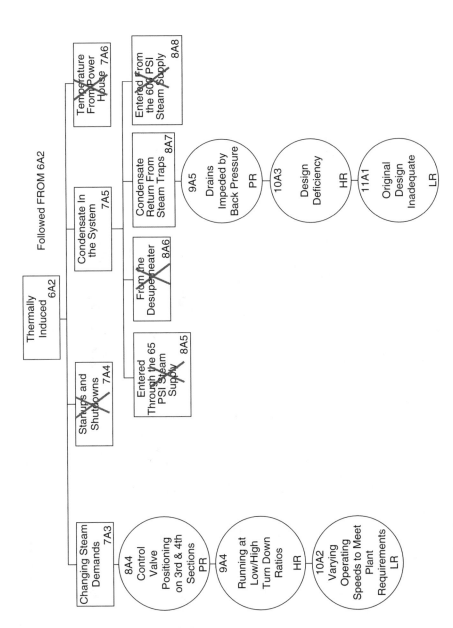

FIGURE 12.21 (continued)

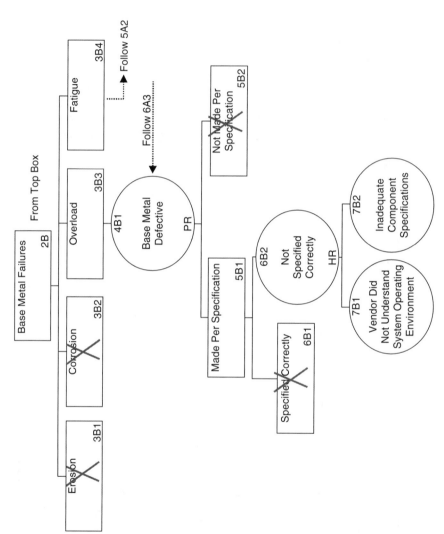

FIGURE 12.21 (continued)

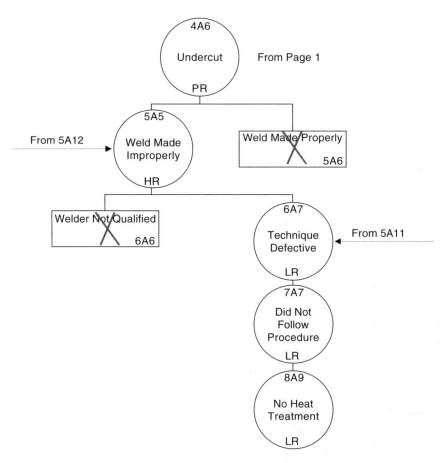

FIGURE 12.21 (continued)

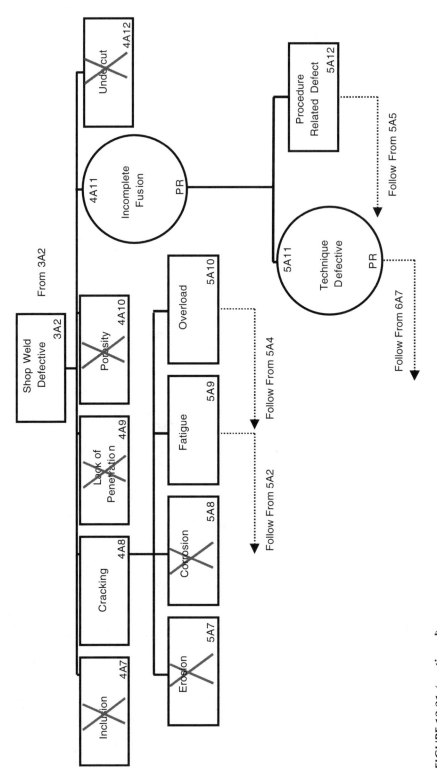

FIGURE 12.21 (continued)

Index

HQ